高等学校教材

复合材料工程技术专业英语

English for Composite Materials Engineering and Technology

主　编　王新玲

副主编　李璋琪　徐滕州

西北工业大学出版社

西　安

【内容简介】 本书共分为七章，内容包括飞机结构(Aircraft Structure)、复合材料简介(Introduction to Composite Materials)、复合材料零件制造的原材料(Raw Materials for Part Fabrication)、复合材料成型工艺(Manufacturing Processing of Composite Materials)、复合材料连接技术(Joining of Composite Materials)、复合材料维修技术(Composites Repair)、复合材料检测技术(Composites Inspection)。每章均由课文(Intensive Reading)、注释(Notes)、练习(Exercises)和阅读材料(Reading Material)、单词和词组(Words and Expressions，包括音标及解释)组成。

本书可用作本科和高职高专复合材料工程技术专业师生教材，也可供企业技术人员参考阅读。

图书在版编目(CIP)数据

复合材料工程技术专业英语 / 王新玲主编. — 西安：西北工业大学出版社，2021.8

ISBN 978-7-5612-7819-2

Ⅰ. ①复… Ⅱ. ①王… Ⅲ. ①复合材料-英语 Ⅳ. ①TB33

中国版本图书馆 CIP 数据核字(2021)第 153527 号

FUHE CAILIAO GONGCHENG JISHU ZHUANYE YINGYU

复合材料工程技术专业英语

责任编辑：张 潼 **策划编辑**：杨 军

责任校对：杨 军 张 炜 **装帧设计**：

出版发行：西北工业大学出版社

通信地址：西安市友谊西路 127 号 邮编：710072

电 话：(029)88491757，88493844

网 址：www.nwpup.com

印 刷 者：陕西向阳印务有限公司

开 本：787 mm×1 092 mm 1/16

印 张：7.75

字 数：203 千字

版 次：2021 年 8 月第 1 版 2021 年 8 月第 1 次印刷

定 价：29.00 元

PREFACE
前　言

复合材料优异的耐腐蚀性、高强度与抗冲击性，使其在航空航天、建筑、防腐、管道、水处理等领域广泛应用。近年来，复合材料的应用领域更加广阔，在汽车、新能源、桥梁建筑等市场大显身手。中国消费市场不断扩大以及全球制造中心向中国转移，Hexcel、Gruit、Vestas、Alcan、Menzolite、Airbus、Samtechap、Huntsman 等先后在国内投资建厂或扩大生产规模，取得了良好的经济和社会效益。这样的时代背景，对学习复合材料专业学生的专业英语水平提出了更高的要求。我们培养的学生未来在企业中不仅技术一流，更可以用英语和国外技术专家进行交流探讨。所以学习复合材料专业的学生应该具备通识性的复合材料专业英语基础。

为了激发学生学习兴趣，本书合理选择和安排教学内容，从介绍飞机主要组成部分引出复合材料简介、成型工艺、连接技术、维修技术和检测技术，内容面面俱到，但也都是点到为止，是一本通识性的复合材料专业书。在版面设计方面，对文中出现的专业词汇在右栏进行了翻译提示，以便学生更好地理解章节内容。每一章节末附有词汇表、音标和翻译；对文中较难理解长句的精准翻译；针对章节内容给出翻译句子、短语、练习题等；补充材料是拓展阅读或者任务单，为学有余力的学生加餐。每课的课文和阅读材料均取自各种不同风格的专业文献资料，可独立成篇，具有词汇量适中，覆盖面广的特点。

全书由王新玲主编，南京工业职业技术学院徐腾州主审并校对。本书 Chapter 1 至 Chapter 5 由王新玲编写，Chapter 6 和 Chapter 7 由李璋琪编写。感谢李彩林、付成龙、吴悦梅等对全稿的审核校订，特别感谢成都航空职业技术学院王郁萱、马鹏凯、李凯等学生对图片的编辑处理。在本书编写的过程中，曾参阅了国内外相关文献资料，在此对其作者一并感谢。

限于笔者水平，虽已尽了最大努力，但书中一定存在一些不足之处，敬请读者将使用中发现的问题及时反馈，以便笔者进一步完善教材。

编　者

2021 年 2 月

CONTENT
目　录

Chapter 1 Aircraft Structure

Warm-up Discussion:

①Do you know the distribution of composite material of aircraft structure?

②Do you know how to describe the function/feature of aircraft parts?

For example, the materials distribution on A380 structure: 60% aluminum, 22% composite material, 10% titanium & steel, 3% glare and 5% other materials. Composite material is mainly used in center wing box wing ribs, outer flaps, floor beams, vertical tail, horizontal tail (as shown in Fig. 1-1).

飞机结构
功能;特性
分配
铝合金;复合材料;钛合金;钢
中央翼盒;翼肋;外襟翼;地板梁;垂直尾翼;水平尾翼

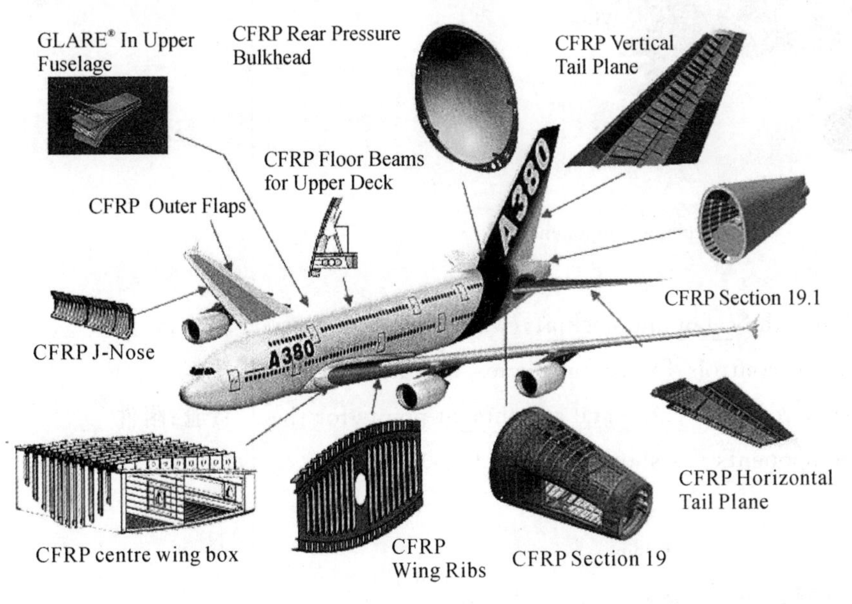

Fig. 1-1 The distribution of composite material of A380 structure

Composite material has become the most important basic material of aircraft structure. In order to know composite material, we should learn about the structure of airplane firstly.

An aircraft is a device that is used, or intended to be used, for flight. We also defines airplane as an engine-driven, fixed-wing aircraft heavier than air that is supported in flight by the dynamic reaction of air against its wings. Categories of aircraft for certification of airmen include airplane, rotorcraft, lighter-than-air, powered-lift and glider. This paragraph provides a brief introduction to the airplane and its major components.

装置
定义;发动机驱动;固定机翼
动力反作用种类
旋翼机;鉴定;动力升降机;滑翔机
组成部分

1. Major Components

Although airplanes are designed for a variety of purposes, most of them have the same major components. The overall characteristics are largely determined by the original design objectives. Most airplane structures include a fuselage, wings, an empennage, landing gear, and a powerplant (as shown in Fig. 1-2).

特征
最初的;目标
机身;机翼;尾翼;
起落架;动力装置

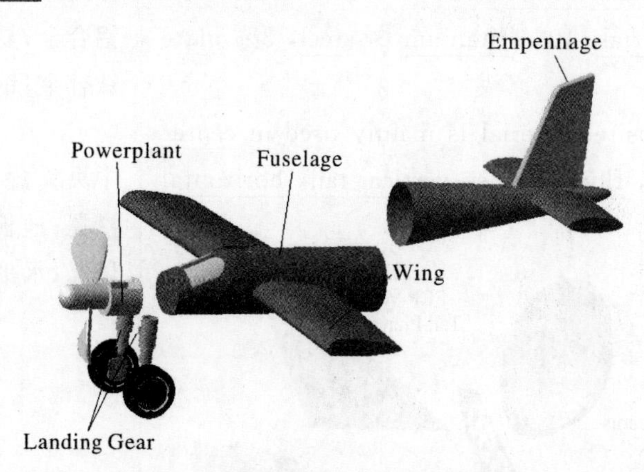

Fig. 1-2 Airplane components

2. Fuselage

The fuselage includes the cabin and cockpit, which contains seats for the occupants and the controls for the airplane. In addition, the fuselage may also provide room for cargo and attachment points for the other major airplane components (as shown in Fig. 1-3).

客舱;驾驶室
乘客
货舱;附件

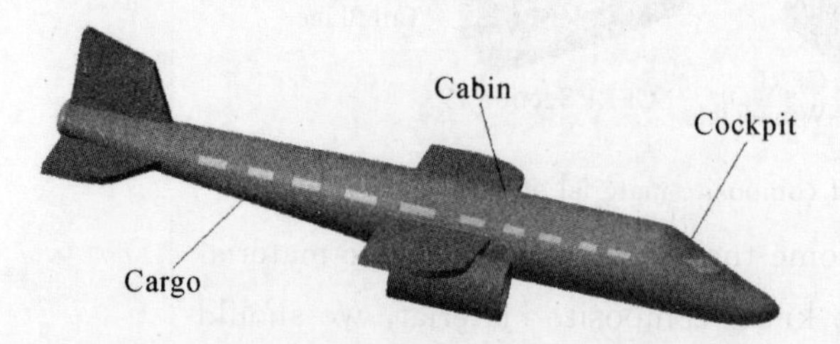

Fig. 1-3 The main components of fuselage

Some aircraft utilize an open truss structure. **The truss-type fuselage is constructed of steel or aluminum tubing. Strength and rigidity is achieved by welding the tubing together into a series of triangular shapes, called trusses** (as shown in Fig. 1-4).

利用;桁架
铝合金管;强度;刚度
焊接三角

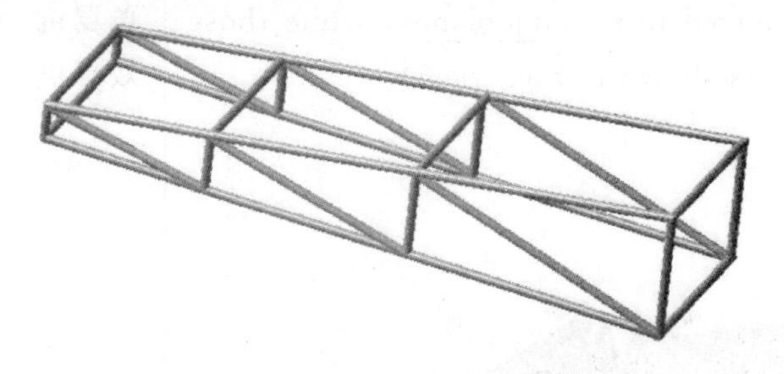

Fig. 1-4 Open truss structure

3. Wings

The wings are airfoils attached to each side of the fuselage and are the main lifting surfaces that support the airplane in flight. There are numerous wing designs, sizes, and shapes used by the various manufacturers. Each fulfills a certain need with respect to expected performance for demand. Wings can control at various operating speeds, the amount of lift generated, balance, and stability. All change as the shape of the wing is altered (as shown in Fig. 1-5).

机翼
升力
许多
完成;性能
运行速度
稳定性
改变

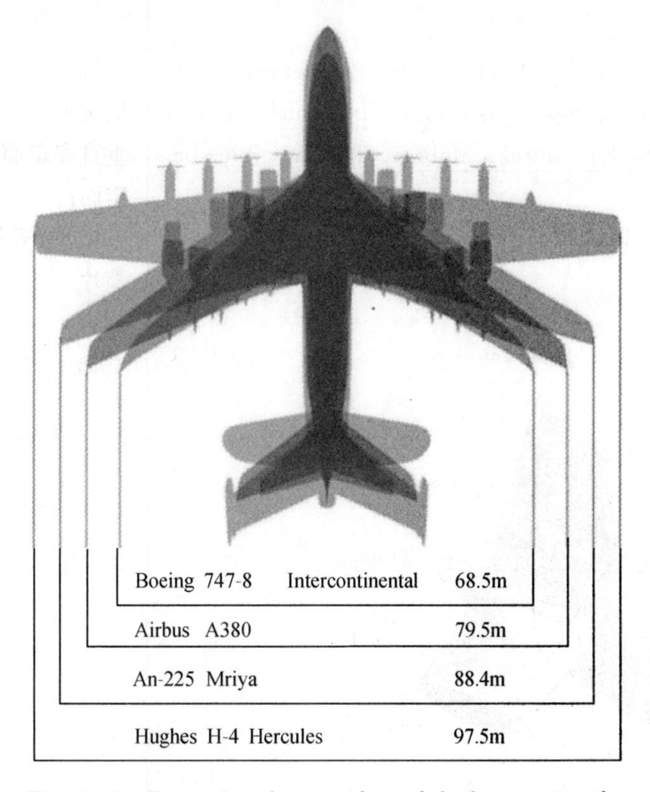

Fig. 1-5 Comparison between four of the largest aircraft

How the wing produces lift is explained in aerodynamics principle? Wings may be attached at the top, middle, or lower portion of the fuselage. These designs are referred to as high-wing, mid-wing, and low-wing, respectively. The number of wings can also vary. Airplanes with a single set of wings are referred to as monoplanes, while those with two sets are called biplanes(as shown in Fig. 1-6).

空气动力学原理
位置
单翼机
双翼机

Fig. 1-6 Monoplane and biplane

4. Empennage

The correct name for the tail section of an airplane is empennage. The empennage includes the entire tail group, consisting of fixed surfaces such as the vertical stabilizer and the horizontal stabilizer. The movable surfaces include the rudder, the elevator, and one or more trim tabs (as shown in Fig. 1-7).

尾部
固定面
垂直安定面;水平安定面;
方向舵;升降舵;调整片

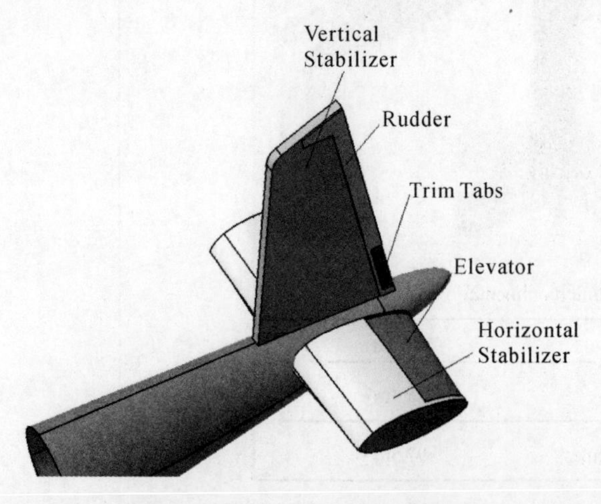

Fig. 1-7 Empennage components

5. Landing Gear

The landing gear is the principle support of the airplane when parked, taxiing, taking off, or when landing. The most common type of landing gear consists of wheels, but airplanes can also be equipped with floats for water operations, or skis for landing on snow.

滑行;起飞

救生衣;滑雪板

The landing gear consists of three wheels, two main wheels and a third wheel positioned either at the front or rear of the airplane. Landing gear employing a rear mounted wheel is called conventional landing gear.

后置的

Airplanes with conventional landing gear are sometimes referred to as tailwheel airplanes. When the third wheel is located on the nose, it is called a nosewheel, and the design is referred to as a tricycle gear. A steerable nosewheel or tailwheel permits the airplane to be controlled throughout all operations while on the ground (as shown in Fig. 1-8).

尾轮

鼻轮;三轮车式

Fig. 1-8 Landing gear

6. The Powerplant

The powerplant usually includes both the engine and the propeller. The primary function of the engine is to provide the power to turn the propeller. It also generates electrical power, provides a vacuum source for some flight instruments, and in most single-engine airplanes, provides a source of heat for the pilot and passengers. The engine is covered by a cowling, or in the case of some airplanes, supported by nacelles(as shown in Fig. 1-9).

发动机;螺旋桨

真空

设备

发动机罩

吊舱

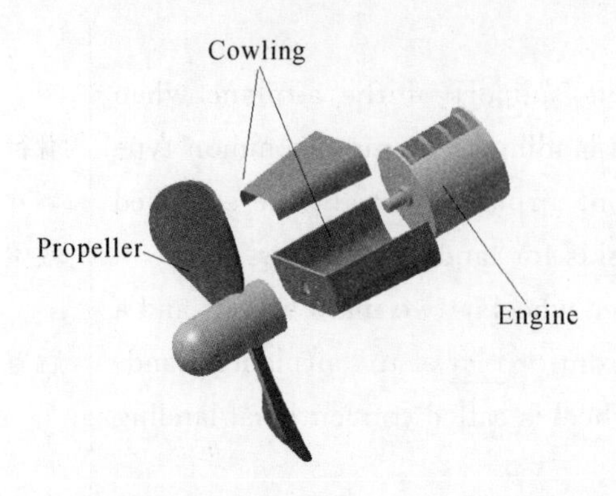

Fig. 1-9 Engine compartment

The purpose of the cowling or nacelle is to streamline the flow of air around the engine and to help cool the engine by ducting air around the cylinders. The propeller, mounted on the front of the engine, translates the rotating force of the engine into a forward acting force called thrust that helps move the airplane through the air.

流线化
用管道输送
汽缸
旋转;推力

This concludes the brief introduction to the main airplane components.

Notes

1. An aircraft is a device that is used, or intended to be used, for flight. We also defines airplane as an engine-driven, fixed-wing aircraft heavier than air that is supported in flight by the dynamic reaction of air against its wings.

飞机是一种用于或将用于飞行的机器。我们还将飞机定义为一种引擎驱动的固定翼飞机,可以靠自身的动力在大气中飞行,且重于空气的航空器。

2. The truss-type fuselage is constructed of steel or aluminum tubing. Strength and rigidity is achieved by welding the tubing together into a series of triangular shapes, called trusses.

桁架式机身由钢管或铝管构成。桁架强度和刚度通过将管道焊接成一系列三角形来实现。

3. Airplanes with conventional landing gear are sometimes referred to as tailwheel airplanes. When the third wheel is located on the nose, it is called a nosewheel, and the design is referred to as a tricycle gear. A steerable nosewheel or tailwheel permits the airplane to be controlled throughout all operations while on the ground.

带有传统起落架的飞机有时被称为尾轮飞机。当第三轮位于前端时,它被称为前轮,这种设计被称为三轮车式。一个可操纵的前轮或尾轮允许飞机在地面上的所有操作都被控制。在地面上的飞机通过可操纵的前轮或尾轮来控制。

4. The purpose of the cowling or nacelle is to streamline the flow of air around the engine and to help cool the engine by ducting air around the cylinders. The propeller, mounted on the front of the engine, translates the rotating force of the engine into a forward

acting force called thrust that helps move the airplane through the air.

整流罩或机舱的目的是使发动机周围的气流流线化，并通过在气缸周围引导空气来帮助发动机冷却。安装在发动机前部的螺旋桨将发动机的旋转力转化为向前的动力，称为推力，帮助飞机在空中飞行。

Exercises

1. Translate the following sentences into Chinese.

(1)Categories of aircraft for certification of airmen include airplane, rotorcraft, lighter-than-air, powered-lift, and glider.

(2)Most airplane structures include a fuselage, wings, an empennage, landing gear, and a powerplant.

(3)The fuselage includes the cabin and cockpit, which contains seats for the occupants and the controls for the airplane.

(4)The wings are airfoils attached to each side of the fuselage and are the main lifting surfaces that support the airplane in flight.

2. Give a definition for each following word.

(1)Aircraft (2)fuselage (3)wings (4)Landing gear

(5)empennage

3. Reading comprehension.

(1) Which is not Category of aircraft for certification of airmen?

A. airplane B. rotorcraft C. balloon D. glider

(2) Which is not the major component of airplane structures?

A. fuselage B. empennage C. powerplant D. toilet

(3) Which is the movable surfaces of an empennage?

A. vertical stabilizer B. rudder C. horizontal stabilizer D. fairing

(4) What is the primary function of the engine?

A. provide the power to turn the propeller

B. generate electrical power

C. provide a vacuum source for some flight instruments

D. provide a source of heat for the pilot and passengers

Reading Material

Competition of composites and metals

The use of high-performance polymer-matrix fiber composites in aircraft structures has grown steadily, although not as dramatically as predicted at that time. This is despite the significant weight-saving and other advantages that these composites can provide.

The main reason for the slower-than-anticipated take-up is the high cost of aircraft components made of composites compared with similar structures made from metal, mainly aluminum alloys. Other factors include the high cost of certification of new components and their relatively low resistance to mechanical damage, low through-thickness strength, and (compared with titanium alloys) temperature limitations. Thus, metals will continue to be favored for many airframe applications. The most important polymer-matrix fiber material and the main subject of this and the previous book, *Composite Materials for Aircraft Structures*, is carbon fiber-reinforced epoxy (carbon/epoxy). Although the raw material costs of this and similar composites will continue to be relatively high, with continuing developments in materials, design and manufacturing technology, their advantages over metals are increasing.

However, competition will be fierce with continuing developments in structural metals. In aluminum alloys developments include improved toughness and corrosion resistance in conventional alloys, new lightweight alloys (such as aluminum lithium), low-cost aerospace-grade castings, mechanical alloying (high-temperature alloys), and super-plastic forming. For titanium, they include use of powder preforms, casting, and super-plastic-forming and diffusion bonding.

However, that airframes (and engines) will continue to be a mix of materials. These will include composites of various types and a range of metal alloys, the balance depending on structural and economic factors (showed in Fig. 1-10).

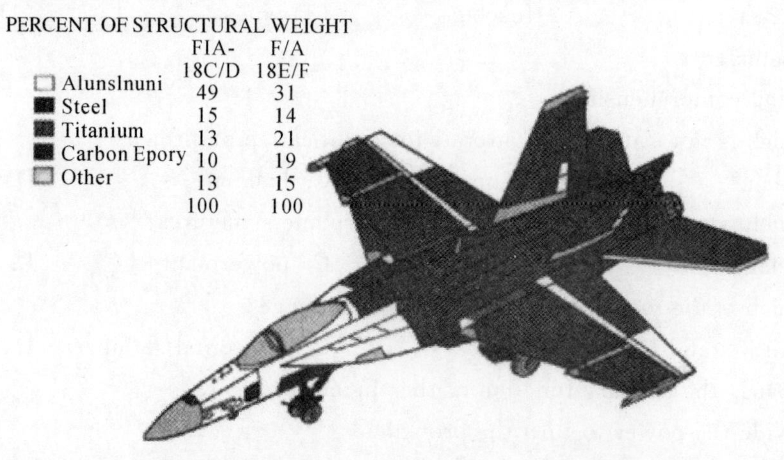

Fig. 1-10 Schematic diagram of fighter aircraft F-18 E/F

(For comparison details of the structure of the earlier C/D model are also provided in the inset table.)

New Words and Expressions

device	[dɪ'vaɪs]	n. 装置;设备
categories	['kætɪgərɪz]	n. 种类,类别
certification	[ˌsɜːrtɪfɪ'keɪʃn]	n. 证明,鉴定,证书
rotorcraft	['routəˌkrɑːft]	n. 旋翼机;旋翼飞机
glider	['glaɪdə]	n. 滑翔机
components	[kəm'pounənt]	n. (机器、设备等的)构成

characteristics	[kærɪktə'rɪstɪk]	n. 性质;特性,特征
original	[ə'rɪdʒənəl]	adj. 原始的;最初的
objectives	[əbd'ʒektɪvz]	n. 目标
fuselage	['fjuːslɑːʒ]	n.(飞机的)机身
wings	[wɪŋz]	n. 机翼
empennage	['empɪnɪdʒ]	n. 尾部,尾翼
powerplant	['pauərplɑːnt]	n. 动力装置
cabin	['kæbɪn]	n. 客舱
cockpit	['kɑːkpɪt]	n. 驾驶员座舱
occupants	['ɒkjuːpənts]	n. 乘客
cargo	['kɑːrgoʊ]	n. 货舱
attachment	[ə'tætʃmənt]	n. 附件,附属物
utilize	['jutlˌaɪz]	vt. 利用,使用
truss	[trʌs]	n. 桁架
steel	[stil]	n. 钢,钢铁;钢制品
aluminum	[ə'ljuːminəm]	n. 铝
tubing	['tuːbɪŋ]	n.(金属、塑料等的)管形材料
strength	[streŋθ]	n. 强度
rigidity	[rɪ'dʒɪdɪti]	n. 刚度
welding	[weldɪŋ]	vt. 熔接;锻接;使结合
triangular	[traɪ'ʃæŋgjələ]	adj. 三角(形)的
airfoils	['eəfɔɪlz]	n. 翼型
lifting	['lɪftɪŋ]	v. 举起,抬起(lift 的现在分词)
fulfills	[foɪ'fɪlz]	v. 履行(诺言);执行(命令)
performance	[pər'fɔːrməns]	n. 性能
monoplane	['mɑːnəpleɪn]	n. 单翼机
biplanesn	['baɪpleɪnz]	n. 双翼飞机,复翼飞机
tail	[tel]	n. 尾;尾部
rudder	['rʌdə]	n. [航]方向舵
elevator	['ɛləˌvetə]	n. 电梯;升降舵
taxiing	[tæk'sɪɪŋ]	v. 滑行达到(目的)使结束
floats	[floʊts]	n. 漂浮物;浮板
skis	[skiːz]	n. 滑雪板(ski 的名词复数)
rear	[rɪə]	n. 后部,面
tailwheel	[teɪl'wiːl]	n. 尾轮

nosewheel	[ˈnouzˌwiːl]	n. 前轮;(飞机)头部机鼻轮
engine	[ˈɛndʒɪn]	n. 发动机,引擎
propeller	[prəˈpɛlə]	n. 螺旋桨,推进器
vacuum	[ˈvækjuəmn]	n. 真空
instruments	[ˈɪnstrumənts]	n. 工具;仪器
cowling	[ˈkaʊlɪŋ]	n. 飞机引擎罩
nacelle	[nəˈsel]	n. 飞机的引擎机舱
streamline	[ˈstrimˌlaɪn]	vt. 把……做成流线型
ducting	[ˈdʌktɪŋ]	n. 管道,导管
cylinders	[ˈsɪlɪndəz]	n. 圆柱;汽缸;圆筒
rotating	[roʊˈteɪtɪŋ]	v. (使某物)旋转(转动)
thrust	[θrʌst]	n. 推力
engine-driven		发动机驱动
fixed-wing		固定机翼的
dynamic reaction		动力反作用
powered-lift		动力升降机
landing gear		起落架装置
aerodynamics pages		空气动力学叶面
fixed surfaces		固定表面
horizontal stabilizer		水平稳定器
trim tabs		调整片
taking off		起飞
tricycle gear		三轮车装置

Chapter 2 Introduction to Composite Materials

Warm-up Discussion:

①Why do you learn about composites?

②Why are the rapid development and use of composite materials beginning in the 1940s?

③What is the principle of combined action of composites?

④What are composite materials?

⑤What are the functions of fibers and matrix of composites? 纤维;基体

A composite, in the present context, is a multiphase material that is artificially made, as opposed to one that occurs or forms naturally. In addition, the constituent phases must be chemically dissimilar and separated by a distinct interface. Thus, most metallic alloys and many ceramics do not fit this definition, because their multiple phases are formedas a consequence of natural phenomena(as shown in Fig. 2-1).

多相的
人工地;自然地
相;截然不同的
相界面;合金
陶瓷
现象

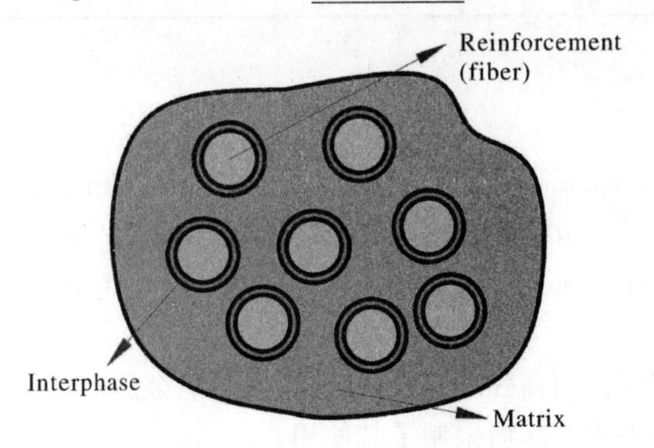

Fig. 2-1 Schematic illustration of composite constituents

Unit 1 History of Composite Materials

The rapid development and use of composite materials beginning in the 1940s had three main driving forces: ① **Military vehicles, such as airplanes, helicopters, placed a premium on high-strength, light-weight**

军用交通工具
直升机;重视

materials. While the metallic components that had been used up to that point certainly did the job in terms of mechanical properties, the heavy weight of such components was prohibitive. The higher the weight of the plane or helicopter itself, the less cargo its engines could carry. ② Polymer industries were quickly growing and tried to expand the market of plastics to a variety of applications. **The emergence of new, light-weight polymers from development laboratories offered a possible solution for a variety of uses, provided something could be done to increase the mechanical properties of plastics.** ③ The extremely high theoretical strength of certain materials, such as glass fibers, was being discovered. The question was how to use these potentially high-strength materials to solve the problems posed by the military's demands. There are four generations of composites. Table 2 - 1 gives a very brief introduction to the development of composite materials.

金属的
禁止
聚合物
扩大；塑料
出现
解决方案
力学的
理论的
潜在地

Table 2 - 1 Four Generations of Composites

Different Generation of Composites	The Major Composites
The First Generation (1940s)	Glass Fiber Reinforced Composites
The Second Generation (1960s)	High Performance Composites in the Post-Sputnik Era
The Third Generation (1970s & 1980s)	The Search for New Markets and the Synergy of Properties
The Fourth Generation (1990s)	Hybrid Materials, Nanocomposites and Biomimetic Strategies

Notes

1. Military vehicles, such as airplanes, helicopters, placed a premium on highstrength, light-weight materials. While the metallic components that had been used up to that point certainly did the job in terms of mechanical properties, the heavy weight of such components was prohibitive.

飞机、直升机等军用飞行器需要使用高强度、轻量化材料。虽然金属部件在机械性能方面确实已经达到了要求，但是这些部件太重从而限制了其使用。

2. The emergence of new, light-weight polymers from development laboratories offered a possible solution for a variety of uses, provided something could be done to increase the mechanical properties of plastics.

来自于研发中心新的轻质聚合物的出现，为各种用途提供了可能的解决方案，提供了提高塑料的机械性能的改进措施。

Exercises

1. Translate the following sentences into Chinese.

(1)The rapid development and use of composite materials beginning in the 1940s had three main driving forces.

(2)The higher the weight of the plane or helicopter itself, the less cargo its engines could carry.

(3)Polymer industries were quickly growing and tried to expand the market of plastics to a variety of applications.

(4)The question was how to use these potentially high-strength materials to solve the problems posed by the military's demands.

2. Reading comprehension.

(1)When was the rapid development and use of composite materials beginning?

A. B. C.　　B. 1940s　　C. 1960s　　D. 1990s

(2)Which is not the major driving forces of development of composites?

A. Military's demands　　B. Polymer industries

C. The high performance fiber was discovered　　D. Sports goods

Reading Material

Origin of composite materials

One of the earliest uses of composite material was by the ancient Mesopotamians around 3400 B. C. , when they glued wood strips at different angles to create plywood. The concept of "composite" building construction has existed since ancient times. Civilizations throughout the world have used basic elements of their surrounding environment in the fabrication of dwellings including mud/straw and wood/clay. "Bricks" were and still are made from mud and straw (as shown in Fig. 2-2).

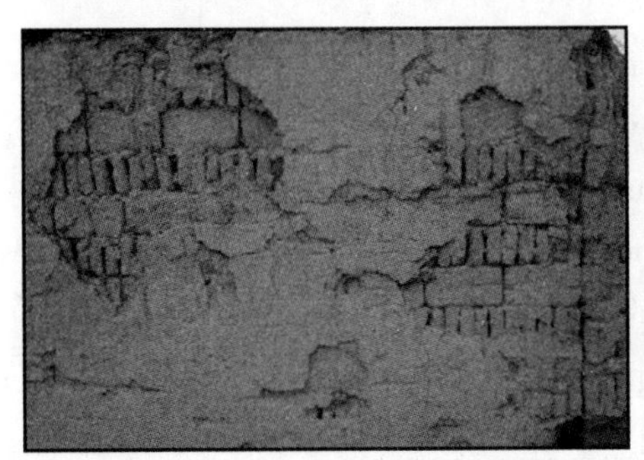

Fig. 2-2 The earliest composite materials

In the 12th century A. D. , Mongol warriors used composite materials (bamboo, silk, cattle tendons and horns, and pine resin) to craft archery bows that were swifter and more powerful than those of their rivals: they put the tendons on the tension (outer) side and sheets of horn on the compression (inner) side of the bow over a core of bamboo. They tightly wrapped the structure with silk and sealed it with pine resin. A museum tested some of the surviving bows, now more than 900 years old, and found that the old bows were nearly as strong as modern ones—and could hit targets as far away as 490 yards (the length of nearly five football fields).

New Words and Expressions

helicopters	[ˈhelɪkɒptəz]	n. 直升机
rocket	[ˈrɑːkɪt]	n. 火箭；火箭发动机；火箭发射器
premium	[ˈprimiəm]	n. 加付款;加价;奖品
metallic	[məˈtælɪk]	adj. 金属的；金属性的；金属制的
prohibitive	[proˈhɪbɪtɪv]	adj. 禁止的；禁止性的；抑制的
polymer	[ˈpɒlɪmə]	n. 聚合物
expand	[ɪkˈspænd]	vt. 扩张；使……变大；详述
plastics	[ˈplæstɪks]	n. 塑料,塑料制品(plastic 的名词复数)
emergence	[ɪˈməːdʒəns]	n. 出现,发生；暴露
solution	[səˈluʃən]	n. 溶液;解决方案
mcchanical	[mɪˈkænɪkəl]	adj. 机械的,机械学的
theoretical	[ˌθiːəˈretɪkl]	adj. 理论的；推想的,假设的
potentially	[pəˈtenʃəlɪ]	adv. 潜在地；可能地

Unit 2 Basic Concepts of Composite Materials

There are more than 50,000 materials available to engineers for the design and manufacturing of products for various applications. It is difficult to study all of these materials individually. Therefore, a broad classification is necessary for simplification and characterization.

可获得的

分类;简单化;特性

These materials, depending on their major characteristics (e. g. stiffness, strength, density, and melting temperature), can be broadly divided into four main categories: metals, polymers, ceramics, and composites.

硬度;强度;密度;熔化温度聚合物;陶瓷制品

Composites, which consist of two or more separate materials combined in a structural unit, are typically made from various combinations of the

混合

other three materials. In the early days of modern man-made composite materials, the constituents were typically macroscopic. **As composites technology advanced over the last few decades, the constituentsmaterials, particularly the reinforcement materials, steadily decreased in size. Most recently, there has been considerable interest in "nanocomposites" having nanometer-sized reinforcements, such as carbon nanotubes.**

Table 2 - 2 depicts the properties of some selected materials in each class in terms of density (specific weight), stiffness, strength, and maximum service temperature.

宏观的
成分
增强体
纳米复合材料
纳米尺寸;碳纳米管
描述
密度
最高使用温度

Table 2 - 2　Typical properties of some engineering materials

Material		Density (ρ) g/cm^3	Tensile Modulus (E) GPa	Tensile Strength (σ) GPa	Specific Modulus (E/ρ) m^2/s^2	Specific Strength (σ/ρ) m^2/s^2	Max Service Temp °C
Metals	Cast iron, grade 20	7.0	100	0.14	14.3	0.02	230 - 300
	Steel, AISI 1045 hot rolled	7.8	205	0.57	36.3	0.073	500 - 650
	Aluminum 2024-T4	2.7	73	0.45	27.0	0.17	150 - 250
	Aluminum 6061-T6	2.7	69	0.27	25.5	0.10	150 - 250
Plastics	Nylon 6/6	1.15	2.9	0.082	2.52	0.071	75 - 100
	Polypropylene	0.9	1.4	0.033	1.55	0.037	50 - 80
	Epoxy	1.25	3.5	0.069	2.8	0.055	80 - 215
	Phenolic	1.35	3.0	0.006	2.22	0.0044	70 - 120
Ceramics	Alumina	3.8	350	0.17	92.1	0.045	1,425 - 1,540
	MgO	3.6	205	0.06	56.9	0.017	900 - 1,000
Short fiber composites	Glass-filled epoxy (35%)	1.90	25	0.30	8.26	0.16	80 - 200
	Glass-filled polyester (35%)	2.00	15.7	0.13	7.25	0.065	80 - 125
	Glass-filled nylon (35%)	1.62	14.5	0.20	8.95	0.12	75 - 110
	Glass-filled nylon (60%)	1.95	21.8	0.29	11.18	0.149	75 - 110
Unidirectional composites	S-glass/epoxy (45%)	1.81	26.3	0.87	21.8	0.48	80 - 215
	Carbon/epoxy (61%)	1.59	27.0	1.73	89.3	1.08	80 - 215
	Kevlar/epoxy (53%)	1.35	25.5	1.1	47.1	0.81	80 - 215

1. What Are Composites?

A composite is a material that is formed by combining two or more materials to achieve some superior properties. Almost all the materials which we see around are composites. Some of them like woods, bones, stones, etc. are natural composites, as they are either grown in nature or developed by natural processes.

较高的;特性

We discussed man-made composites. **A composite is a multiphase material that is artificially made, as opposed to one that occurs or forms naturally. In addition, the constituent and disconstituent phases must be chemically dissimilar and separated by a distinct interface.**

多相

人工地

明显的;界面相

Typically, composite material is formed by reinforcing fibers in a matrix resin as shown in Fig. 2-3. The reinforcements can be fibers, particulates or whiskers, and the matrix materials can be metals, plastics, or ceramics.

增强材料

基体树脂

颗粒;晶须

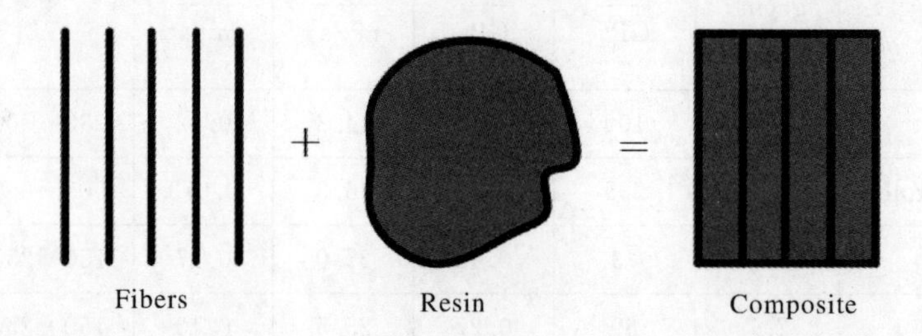

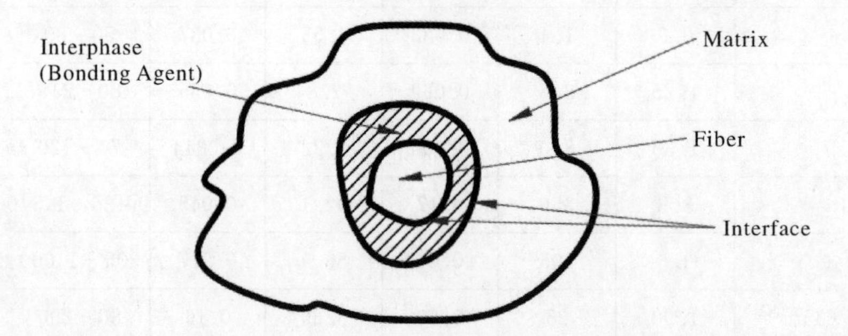

Fig. 2-3 Formation of a composite materials

2. Classification of Composites

Composites materials are usually classified according to the type of reinforcement used. Three broad classes of composites are particulate composite, unidirectional discontinuous fibers composite and unidirectional continuous fibers composites. Each has unique properties and application potential (as shown in Fig. 2-4).

颗粒

单向的;非连续

连续的

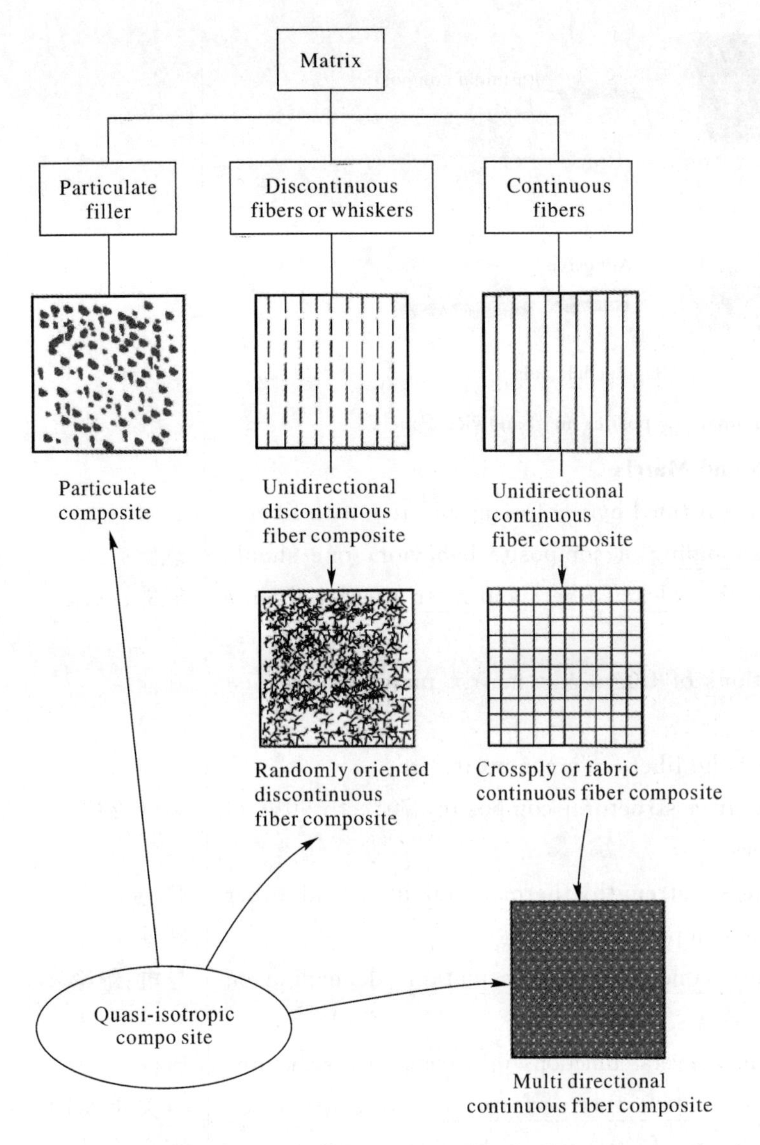

Fig. 2－4 Classification of composites

Structural composites can be broadly subdivided into two main categories: laminar composites and sandwich panels(as shown in Fig. 2－5). A laminar composite is composed of two-dimensional sheets or panels that have a preferred high-strength direction aligned fiber-reinforced plastics. Sandwich panels considered to be a class of structural composites, consist of two strong outer sheets, or faces, separated by a layer of less dense material or core, which has lower stiffness and lower strength.

结构复合材料
层状复合材料;夹层板;尺寸的;片材面板;排列
夹层板
层;密集的;芯子

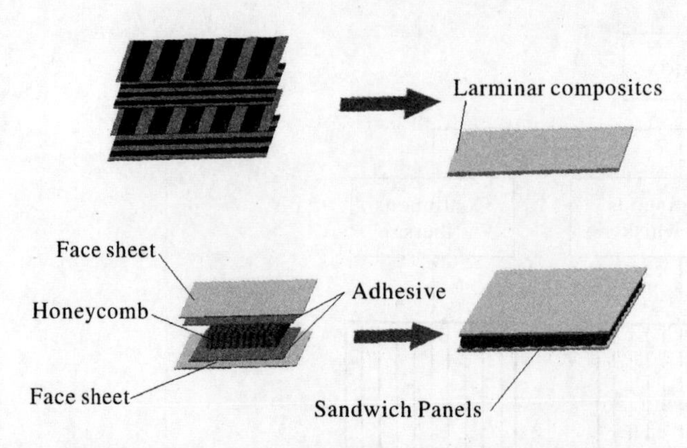

Fig. 2-5 Laminar composites and Sandwich Panels

3. Functions of Fibers and Matrix

A composite material is formed by reinforcing plastics with fibers. To develop a good understanding of composite behavior, one should have a good knowledge of the roles of fibers and matrix materials in a composite. | 理解 职能;基体

The important functions of fibers and matrix materials are discussed below. | 功能

The main functions of the fibers in a composite are:

①To carry the load. In a structural composite, 70% to 90% of the load is carried by fibers. | 承受;载荷

② To provide stiffness, strength, thermal stability, and other structural properties in the composites. | 热稳定性 特性

③ To provide electrical conductivity or insulation, depending on the type of fiber used. | 导电;绝缘

A matrix material fulfills several functions in a composite structure, most of which are vital to the satisfactory performance of the structure. Fibers in and of themselves are of little use without the presence of a matrix material or binder. The important functions of a matrix material include the following: | 履行 至关重要的;令人满意的;性能 胶黏剂

①The matrix material binds the fibers together and transfers the load to the fibers. It provides rigidity and shape to the structure. | 黏结;传输 刚度

②The matrix isolates the fibers so that individual fibers can act separately. This stops or slows the propagation of a crack. | 隔离 传播;破裂

③The matrix provides a good surface finish quality. | 质量

④ The matrix provides protection to reinforcing fibers against chemical attack and mechanical damage (wear). | 保护 侵蚀;磨损

⑤ Depending on the matrix material selected, performance characteristics such as ductility, impact strength, etc. are also influenced. | 性能 延展性;冲击

A ductile matrix will increase the toughness of the structure. For higher toughness requirements, thermoplastic-based composites are selected.

⑥Failure mode is strongly affected by the type of matrix material used in the composite as well as its compatibility with the tiber.

4. Composites Advantages

A distinct advantage of composites, over other materials, is the flexibility of design. By using many combinations of resins and reinforcements, one can design a composite to meet specific strength requirement. Advanced composites for aerospace industry are thus tailored to perform a specific set of functions in a specific environment. Composites opened up the era of materials by design when high modulus continuous fibers such as aramid or carbon fibers were introduced. They are not randomly oriented like short fibers, but carefully aligned into a unidirectional tape. Making such composites is a laborious process with many steps, the end product is more expensive than the standard materials used in mass production. However they offer higher performances for a specific use.

(1)Composite materials have a high specific stiffness(stiffness-to-density ratio).

(2)The specific strength(strength-to-density ratio) of a composite material is very high.

(3)The fatigue strength(endurance limit) is much higher for composite materials.

(4)Composite materials offer high corrosion resistance.

(5)Composite materials offer increased amounts of design flexibility.

(6)Compositematerials can manufacture structures and eliminate joints.

(7)Composites offer good impact properties.

5. Drawbacks of Composites

Although composite materials offer many benefits, they suffer from the following disadvantages:

(1) The materials cost for composite materials is very high compared to that of steel and aluminum.

(2) In the past, composite materials have been used for the fabrication of large structures at low volume (one to three parts per day). The lack of high-volume production methods limits the widespread use of composite materials.

(3)The temperature resistance of composite parts depends on the temperature resistance of the matrix materials.

(4)Composites absorb moisture, which affects the properties and

韧性
基于热塑性的
失效
相容性
灵活性
设计;组合
比强度;先进的
专门制作;执行
特定的
模量;芳纶
随机定向
排列
单向带;劳动密集的
标准
比模量
比强度
疲劳强度
耐腐蚀性
设计的灵活性
减少连接
冲击性能
好处
缺点
成本
铝
制造;产量
普遍的
耐受力
基体
潮气

dimensional stability of the composites. 尺寸

6. Composites Product Fabrication

Composite products are fabricated by transforming the raw material into final shape using one of the manufacturing processes. The Fig. 2－6 shows the various manufacturing processes frequently used in the fabrication of thermoset and thermoplastic composites.

制造 / 制造工艺 / 热固性；热塑性

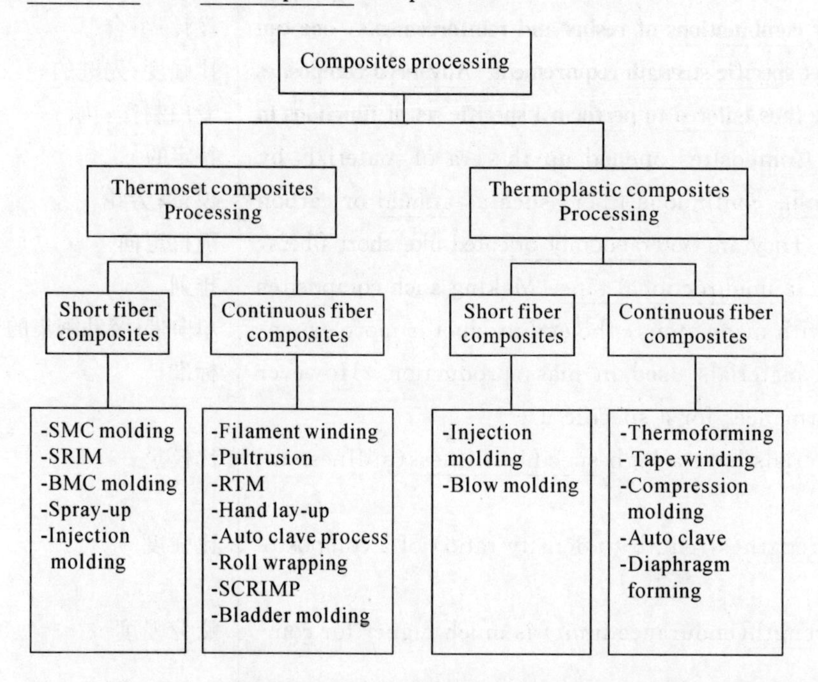

Fig. 2－6　Classification of composites processing technique

7. Composites Markets

There are many reasons for the growth in composite applications, but the primary factor is that the products fabricated by composites are stronger and lighter. 主要的

Broadly speaking, the composites market can be divided into the following industry categories: aerospace, automotive, construction, marine, corrosion resistant equipment, consumer products, appliance/business equipment. 类别；建筑 设备

The aerospace industry was among the first to realize the benefits of composite materials. Airplanes, rockets, and missiles all fly higher, faster, and farther with the help of composites. Glass, carbon, and Kevlar fiber composites have been routinely designed and manufactured for aerospace parts. The aerospace industry primarily uses carbon fiber composites because of their high performance characteristics. 火箭；导弹 / 性能；特征

The hand lay-up technique is a common manufacturing method for the fabrication of aerospace parts; RTM and filament winding are also 手糊成型 / 缠绕成型

being used. The composite components used in the fighter planes are horizontal and vertical stabilizers, wing skins, fin boxes, fiaps, and various other structural components as shown in Table 2 - 3.

战斗机
机翼蒙皮;翼盒;襟翼

Table 2 - 3 Composite components in aircraft applications

Composite Components	
F - 14	Doors, horizontal tails, fairings, stabilizer skins
F - 15	Fins, rudders, vertical tails, horizontal tails, speed brakes, stabilizer skins
F - 16	Vertical and horizontal tails, fin leading edge, skins on vertical fin box
B - 1	Doors, vertical and horizontal tails, fiaps, slats, inlets
AV - 8B	Doors, rudders, vertical and horizontal tails, ailerons, flaps, fin box, fairings
Boeing 737	Spoilers, horizontal stabilizers, wings
Boeing 757	Doors, rudders, elevators, ailerons, spoilers, flaps, fairings
Boeing 767	Doors, rudders, elevators, ailerons, spoilers, fairings

Notes

1. As composites technology advanced over the last few decades, the constituents materials, particularly the reinforcement materials, steadily decreased in size. Most recently, there has been considerable interest in "nanocomposites" having nanometer-sized reinforcements, such as carbon nanotubes.

过去几十年随着复合材料技术的进步,组成材料,特别是增强材料的尺寸稳步减小。最近,人们对具有纳米尺寸增强材料(如碳纳米管)的"纳米复合材料"非常感兴趣。

2. A composite is a multiphase material that is artificially made, as opposed to one that occurs or forms naturally. In addition, the constituent and disconstituent phases must be chemically dissimilar and separated by a distinct interface.

复合材料是人工制造的多相材料,而不是自然形成的材料。此外,连续相和非连续相必须是化学性质截然不同的材料,并由一个明显的界面相分开。

3. A distinct advantage of composites, over other materials, is the flexibility of design. By using many combinations of resins and reinforcements, one can design a composite to meet specific strength requirements.

与其他材料相比,复合材料的一个明显优势是设计的灵活性。通过使用多种不同的树脂和增强材料组合,可以设计出满足比强度要求的复合材料。

4. Advanced composites for aerospace industry are thus tailored to perform a specific set of functions in a specific environment.

量身定做用于航空航天工业的先进复合材料由此可以在特定环境中执行特定的功能。

Exercises

1. Translate the following sentences into Chinese.

(1) A composite is a material that is formed by combining two or more materials to achieve some superior properties.

(2) Structural composites can be broadly subdivided into two main categories: laminar composites and sandwich panels.

(3) To carry the load. In a structural composite, 70% to 90% of the load is carried by fibers.

(4) To provide stiffness, strength, thermal stability, and other structural properties in the composites.

(5) To provide electrical conductivity or insulation, depending on the type of fiber used.

(6) The aerospace industry primarily uses carbon fiber composites because of their high performance characteristics.

2. Give a definition for each following word.

①composites　　②Fibers　　③Matrix

3. Reading comprehension.

(1) Which material is the man-made material?

A. pottery　　B. wood　　C. horn　　D. stone

(2) What is the major advantage of using fibre reinforced plastics (FRP) instead of metals?

A. Low density　　B. higher specific modulus

C. strength properties　　D. weight efficient design

(3) Which type of material isn't the basic categories of the structural materals?

A. metals　　B. composite　　C. polymer　　D. ceramics

(4) Composites are generally used because they have desirable properties that cannot be achieved by any of the constituent materials acting alone. Which is the best reinforcement for composite materials?

A. fiber　　B. Particle　　C. flake　　D. Rod

(5) Which is the strongest and stiffest reinforcement material in existence?

A. whiskers in the micrometer range　　B. fibrous cellulose

C. carbon nanotubes in the nanometer range　　D. glass rods

4. Questions.

(1) What are the different categories of materials? Rank these materials based on density, specific stiffness, and specific strength.

(2) What are the benefits of using composite materials?

(3) What is the function of a matrix in a composite material?

(4) What are the processing techniques for short fiber thermoset composites?

(5) What are the manufacturing techniques available for continuous thermoplastic composites?

Reading Material

What are composites made of?

Composites, also known as Fiber-Reinforced Polymer (FRP) composites, are made from a polymer matrix that is reinforced with an engineered, man-made or natural fiber (like glass, carbon or aramid) or other reinforcing material. The matrix protects the fibers from environmental and external damage and transfers the load between the fibers. The fibers, in turn, provide strength and stiffness to reinforce the matrix—and help it resist cracks and fractures.

In many of our industry's products, polyester resin is the matrix and glass fiber is the reinforcement. But many combinations of resins and reinforcements are used in composites—and each material contributes to the unique properties of the finished product. Fiber, powerful but brittle, provides strength and stiffness, while more flexible resin provides shape and protects the fiber. FRP composites may also contain fillers, additives, core materials or surface finishes designed to improve the manufacturing process, appearance and performance of the final product (as shown in Fig. 2-7).

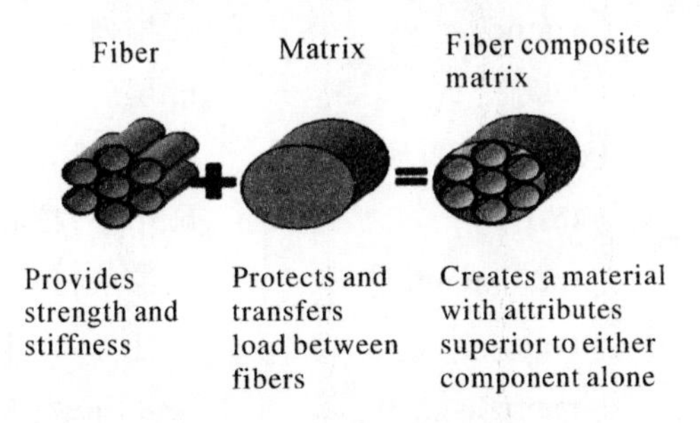

Fig. 2-7 Composition of composite materials

New Words and Expressions

available	[ə'veləbəl]	adj. 可获得的;能找到的
classification	[ˌklæsɪfɪ'keɪʃn]	n. 分类
simplification	[ˌsɪmplɪfɪ'keɪʃn]	n. 单纯化;简单化;简化
characterization	[ˌkærəktəraɪ'zeɪʃn]	n. 特性描述
stiffness	['stɪfnɪs]	n. 硬度
density	['dɛnsɪti]	n. 密度;比重
ceramics	[sə'ræmɪks]	n. 陶瓷制品;陶瓷器
macroscopic	[ˌmækrə'skɒpɪk]	adj. 宏观的;肉眼可见的
constituents	[kən'strtʊrnt]	n. 成分;构成部分
reinforcement	[ˌri:m'fɔ:rsmənt]	n. 增强体;增强相
nanocomposites	[ˌmænə(ʊ)'kɔmpəzɪtz]	n. 纳米复合材料

depict	[dɪˈpɪkt]	vt. 描述;描绘,描画
superior	[suːˈpɪriə(r)]	adj. 较高的;较好的
properties	[pˈrɒpətɪz]	n. 特性
multiphase	[ˈmʌltɪˌfeɪz]	n. 多相
artificially	[ˌɑrtəˈfɪʃəlɪ]	adv. 人为地;人工地
distinct	[dɪˈstɪŋkt]	adj. 明显的,清楚的
interface	[ˈɪntərfeɪs]	n. 界面相,界面层
particulates	[peˈtɪkjələts]	n. 微粒,颗粒
whiskers	[ˌwɪskəs]	n. 晶须
unidirectional	[ˌjuːnɪdəˈrekʃənəl]	adj. 单向的,单向性的
discontinuous	[ˌdɪskənˈtɪnjuəs]	adj. 非连续;不连续的,间断的
continuous	[kənˈtɪnjuəs]	adj. 连续的;延伸的
dimensional	[dɪˈmɛnʃənlː]	adj. 维的;尺寸的
sheets	[ʃiːts]	n. 片材;片料;薄板;板材
panels	[ˈpænlz]	n. 面板
aligned	[əˈlaɪnd]	v. 排列
layer	[ˈleɪə]	n. 层
dense	[dɛns]	adj. 密集的,稠密的;浓密的
core	[kɔːr]	n. 中心;核心
understanding	[ʌndəˈstændɪŋ]	n. 谅解,理解;理解力;协议
roles	[ˈrəʊlz]	n. 任务,作用
matrix	[ˈmeɪtrɪks]	n. 矩阵;发源地;基质
functions	[ˈfʌŋkʃən]	n. 函数;功能
carry	[ˈkærɪ]	vt. 支撑;携带;输送;运载
load	[ləʊd]	n. 载荷
properties	[ˈprɔpətɪz]	n. 特性
conductivity	[kɒndʌkˈtɪvɪtɪ]	n. 导电性;传导性
insulation	[ɪnsjʊˈleɪʃ(ə)n]	n. 隔离,孤立;绝缘;绝缘或隔热的材料
vital	[ˈvaɪt(ə)l]	adj. 至关重要的
binder	[ˈbaɪndə]	n. 缚者;活页封面;黏合剂
transfer	[ˈtrænsfɜ]	n. 传输迁移
		v. 转移;变换;调任
rigidity	[rɪˈdʒɪdətɪ]	n. 坚硬;严格;刚直;死板
isolate	[ˈaisəleit]	n. 孤立者
propagation	[ˌprɒpəˈgeɪʃən]	n. 传播;繁殖;增殖

crack	[kræk]	n. 破裂
quality	[ˈkwɒlətɪ]	n. 质量,品质;特性 adj. 高品质的
protection	[prəˈtekʃ(ə)n]	n. 保护;防卫;护照
attack	[əˈtæk]	vt. &vi. 攻击,进攻,抨击
damage	[ˈdæmɪdʒ]	n. 损害;损失;毁坏 vt. 损害;毁坏
performance	[pəˈfɔːm(ə)ns]	n. 性能;绩效;表演;执行;表现
ductility	[dʌkˈtɪləti]	n. 延展性;柔软性;顺从
impact	[ɪmˈpækt]	n. 冲击,撞击;冲击力 vt. 挤入,压紧
toughness	[ˈtʌfnɪs]	n. 韧性;坚韧;刚性;健壮性
failure	[ˈfeljə]	n. 失败;故障;失败者;破产
compatibility	[kəmˌpætəˈbɪləti]	n. 和谐共处;协调;兼容
flexibility	[ˌflɛksəˈbɪləti]	n. 灵活性;弹性;适应性;柔韧性
design	[dɪˈzaɪn]	n. 设计;图样;图案;目的 v. 设计
combination	[kɒmbɪˈneɪʃz]	n. 组合
advanced	[ədˈvænst]	adj. 先进的;高级的
tailored	[ˈteləd]	adj. 定做的;裁缝做的;剪裁讲究的
perform	[pəˈfɔrm]	v. 执行;履行;表演;运转;举行
specific	[spɪˈsɪfɪk]	adj. 明确的;特殊
modulus	[ˈmɑdʒələs]	n. 系数;模数
aramid	[ˈæræmɪd]	n. 芳香族聚酰胺
aligned	[əˈlaɪn]	adj. 对齐的;均衡的
laborious	[ləˈbɔrɪəs]	adj. 艰苦的;费劲的
standard	[ˈstændəd]	n. 标准;水准 adj. 标准的;合规格的
benefits	[ˈbɛnəfɪt]	n. 津贴费;利益;奖金;救济金;益处
disadvantages	[ˌdɪsədˈvæntɪdʒs]	n. 劣势;不利条件;损害
cost	[kɔst]	vt. 花费;使付出 n. 费用,代价,成本
aluminum	[əˈlʊmɪnəm]	n. 铝
fabrication	[ˌfæbrɪˈkeʃən]	n. 织物;布;组织;构造;建筑物
volume	[ˈvɑljum]	n. 卷;体积;册;音量;容量

widespread	[ˈwʌɪdsprɛd]	adj. 分布广的;普遍的;广泛应用
resistance	[rɪˈzɪstəns]	n. 抵抗;阻力;抗力;电阻
matrix	[metrɪks]	n. 矩阵;发源地;基质
moisture	[ˈmɔɪstʃə]	n. 水分;湿度;潮湿;降雨量
dimensional	[dɪˈmenʃənəl]	adj. 维度的;空间的;尺寸的
fabricated	[ˈfæbrɪket]	v. 制造,组装;伪造,捏造
thermoset	[ˈθəːməuset]	n. 热固性
thermoplastic	[ˌθɜːmoˈplæstɪk]	adj. 热塑性的 n. 热塑性塑料
military vehicles		军用车辆
carbon nanotubes		碳纳米管
specific weight		比强度
maximum service temperature		最高使用温度
reinforcing fibers		增强材料
matrix resin		基体树脂
structural composites		结构复合材料
laminar composites		层状复合材料
sandwich panels		夹层板
randomly oriented		随机取向
thermal stability		热稳定性
satisfactory performance		至关重要的
specific stiffness		比模量
thermoplastic-based		基于热塑性的
unidirectional tape		单向胶带
fatigue strength		疲劳强度
corrosion resistance		耐腐蚀性
design flexibility		设计的灵活性
eliminate joints		整体成型,减少连接
impact properties		冲击性能
manufacturing process		制造过程

Chapter 3 Raw Materials for Part Fabrication

Warm-up Discussion:

①Do you have a good understanding of the function of Matrix and Reinforcement? 基体 增强体

②Which are the common resins and fibers? 树脂;纤维

③Do you know the difference of thermoset and thermoplastic resins? 热固性;热塑性

It is important to have a good knowledge of the various raw materials available for the manufacture of good composite products. Fig. 3-1 depicts the various types of composite manufacturing techniques and their corresponding material systems. In general, raw materials for composite manufacturing processes can be divided into two categories: thermoset-based and thermoplastic-based composite materials. Thermoset plastics are those that once solidified (cured) cannot be remelted. Thermoplastics can be remelted and reshaped once they have solidified. Thermosets and thermoplastics have their own advantages and disadvantages in terms of processing, cost, storage, recyclability and performance. In all these composite systems, there are two major ingredients: reinforcements and matrices.

未加工的

复合材料制造技术
相应的

热固性基的;热塑性基的;被固化;再熔融

存储;再循环能力

成分

Unit 1 Matrices

As discussed, composites are made of reinforcing fibers and matrix materials. Matrix surrounds the fibers and thus protects those fibers against environmental and chemical attack. For fibers to carry maximum load, the matrix must have a lower modulus and greater elongation than the reinforcement. A brief description of the various matrix materials, polymers, metals, and ceramics, is given in this unit. We emphasize the characteristics that are relevant to composites.

增强纤维
基体材料;包裹
侵蚀
模量
延伸

相关的

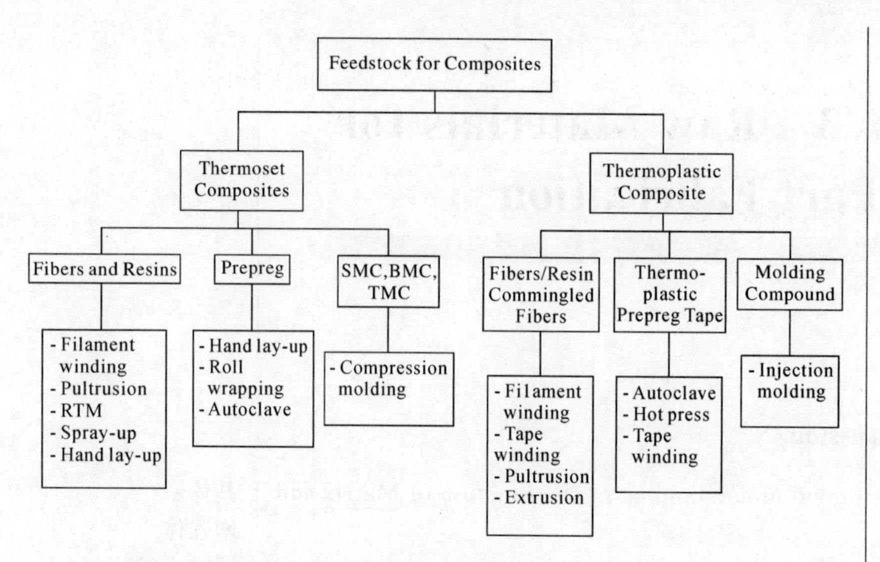

Fig. 3-1 Classification of raw materials

1. Polymers Matrix

Polymer are structurally much complex than metals or ceramics. They are cheap and can be easily processed. On the other hand, polymers have lower strength and modulus and lower temperature use limits. Prolonged exposure to ultraviolet light and some solvents can cause the degradation of polymer properties. Because of predominantly covalent bounding, polymers are generally poor conduction of heat and electricity. Polymers, however, are generally more resistant to chemicals than are metals. Structurally, polymers are giant chainlike molecules (hence the name macromolecules) with covalently bonded carbon atoms forming the backbone of the chain.

加工的
持续很久的
下降
共价键;传导
巨大的
骨架

There area number of methods of classifying polymers. One is to adopt the approach of using their response to thermal treatment and to divide them into thermoplastics and thermosets. Thermoplastics are polymers which melt when heated and resolidify when cooled, while thermosets are those which do not melt when heated but, at sufficiently high temperatures, decompose irreversibly. This system has the benefit that there is a useful chemical distinction between the two groups. Thermoplastics comprise essentially linear or lightly branched polymer molecules, while thermosets are substantially crosslinked materials, consisting of an extensive three-dimensional network of covalent chemical bonding (as shown in Fig. 3-2).

处理
分解
差别
包括
交联的
三维立体

Another classification system, first suggested by Carothers in 1929, is based on the nature of the chemical reactions employed in the polymerization. Here the two major groups are the condensation and

聚合;缩聚

the addition polymers. Condensation polymers are those prepared from monomers where reaction is accompanied by the loss of a small molecule, usually of water. By contrast, addition polymers are those formed by the addition reaction of an unsaturated monomer, such as takes place in the polymerization of vinyl chloride as shown in Fig. 3 - 3.

加成
单体;损失
不饱和的
氯乙烯

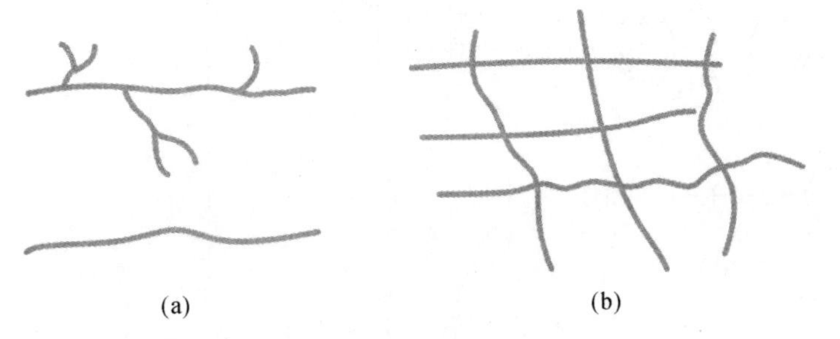

(a) (b)

Fig. 3 - 2 Polymer molecules

(a) linear or lightly branched polymer molecules;

(b) a three-dimensional network polymermolecules

$$nH_2N(CH_2)_6NH_2 + NHOOC(CH_2)_4COOH$$

$$\longrightarrow H\left(NH(CH_2)_6NHCO(CH_2)_4CO\right)_n OH + (2n-1)H_2O$$

$$nCH_2{=}\underset{\displaystyle CL}{\underset{|}{CH}} \longrightarrow \left(CH_2 - \underset{\displaystyle CL}{\underset{|}{CH}}\right)_n$$

Fig. 3 - 3 The condensation reaction and the addition reaction

Resin is a generic term used to designate the polymer, polymer precursor material, and/or mixture or formulation. The common thermoset resins are showed in Fable 3 - 1.

一般的;指定

Table 3 - 1 Typical unfilled thermosetting resin properties

Resin Material	**Density** g/cm^3	**Tensile Modulus** GPa (10^6 psi)①	**Tensile Strength** MPa (10^3 psi)
Epoxy	1.2 - 1.4	2.5 - 5.0(0.36 - 0.72)	50 - 110 (7.2 - 16)
Phenolic	1.2 - 1.4	2.7 - 4.1 (0.4 - 0.6)	35 - 60 (5 - 9)
Polyester	1.1 - 1.4	1.6 - 4.1 (0.23 - 0.6)	35 - 95 (5.0 - 13.8)

(1) Epoxy.

The term epoxy is a general description of a family of polymers which are based on molecules that contain epoxide groups. An epoxide

① 1 psi=6.895kPa.

group is an oxirane structure, a three-member ring with one oxygen and two carbon atoms (shown in Fig. 3-4). Epoxy is a very versatile resin system, allowing for a broad range of properties and processing capabilities. It exhibits low shrinkage as well as excellent adhesion to a variety of substrate materials.

环氧乙烷
通用的
收缩率;胶接
基底

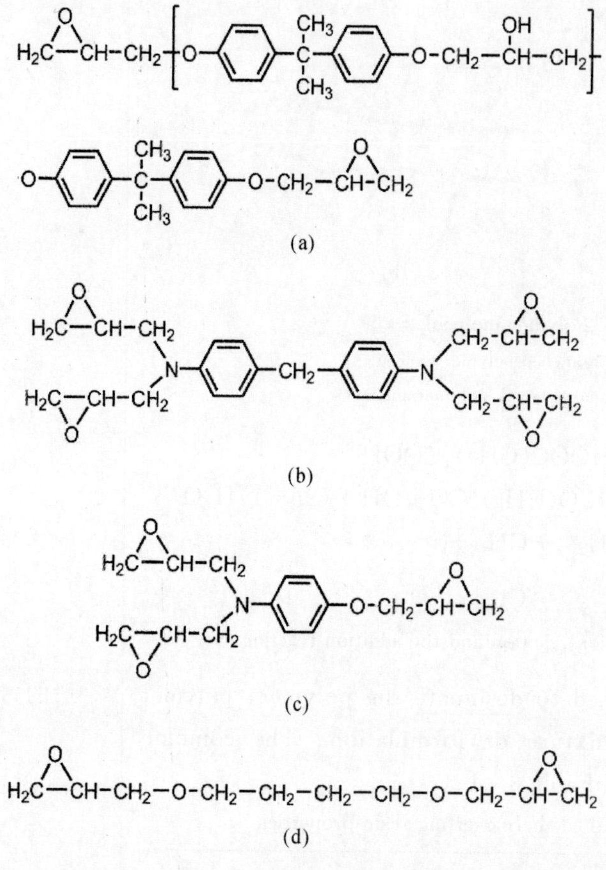

(a)

(b)

(c)

(d)

Fig. 3-4 Major epoxy resins used in aerospace composite matrices:
(a) bisphenol A epichorohydrin (DGEBA) resins;
(b) tetraglycidyl derivative of diamino diphenyl methane (TGGM);
(c) triglycidyl derivative of p-aminophenol (TGAP);
(d) reactive diluent epoxy resin such as the bis epoxy from butane diol.

双酚A树脂
二氨基二苯甲烷四缩水甘油衍生物;
对氨基苯酚的三缩水甘油衍生物;
活性稀释;丁烷二醇

Epoxies are the most widely used in resin materials and used in many applications, from aerospace to sporting goods. Epoxies are used in resins for prepregs and structural adhesives. **The advantages of epoxies are high strength and modulus, low levels of volatiles, excellent adhesion, low shrinkage, good chemical resistance, and ease of processing. Their major disadvantages are brittleness and the reduction of properties in the presence of moisture.** The processing of curing of ep-

预浸料;黏合剂
挥发份
脆性
潮气

oxies is slower than polyester resins. There are varying grades of epoxies with varying levels of performance to meet different application needs. They can be formulated with other materials or can be mixed with other epoxies to meet a specific performance need.

The cost of the resin is also higher than the polyesters. Epoxy-based composites provide good performance at room and elevated temperatures. Curing temperatures vary from room temperature to approximately 350°F (180℃). **The most common cure temperatures range between 250°F and 350°F (120℃ and 180℃). The use temperatures of the cured structure will also vary with the cure temperature.** Higher temperature cures generally yield greater temperature resistance. Cure pressures are generally considered as low pressure molding from vacuum to approximately 100 psi (700 kPa).

Epoxies come in liquid, solid, and semi-solid forms. Liquid epoxies are used in RTM, filament winding, pultrusion, hand lay-up, and other processes with various reinforcing fibers such as glass, carbon, aramid, boron, etc. Semi-solid epoxies are used in prepreg for vacuum bagging and autoclave processes. Solid epoxy capsules are used for bonding purposes. Epoxies are more costly than polyester and vinylesters and are therefore not used in cost-sensitive markets (e. g. automotive and marine) unless specific performance is needed.

(2) Phenolics.

Phenolics are used for aircraft interiors, stow bins and galley walls, as well as other commercial markets that require low-cost, flame-resistant, and low smoke products. It is characterized by both chemical and thermal resistance as well as hardness, and low smoke and toxic degradation products.

The phenolic polymers, often called either phenolic resole or novolacs resins are condensation polymers based upon either a reaction of formaldehyde with a base catalyst and phenol (resole), or a reaction of excess phenol with an acidic catalyst and formaldehyde (novolac). The basic difference between resoles and novolacs consist of no methylol groups in the novolacs and the resulting need for an extension agent of paraformaldehyde, hexamethylenetetramine, or additional formaldehyde as a curative (shown in Fig. 3-5).

Phenolics are generally dark in color and therefore used for applications in which color does not matter.

聚酯
配方
特殊性
升高的
固化
生产
半固态
芳纶；硼
胶囊
乙烯基酯；成本敏感
储物箱；厨房
耐火的
有毒的；降解
酚醛树脂；缩聚
甲醛；苯酚
酸的
羟甲基；六亚甲基四胺

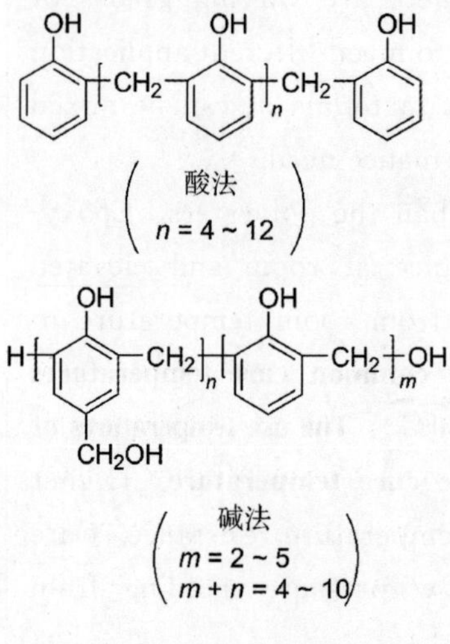

酸法

$n=4\sim12$

碱法

$m=2\sim5$

$m+n=4\sim10$

Fig. 3-5 Major phenolic polymer

These resins have higher molecular weights and viscosities than either parent material. Consequently, they are optimal for processing parts of unusual conformations and complex curvature. The resins allow either press or autoclave cure and allow relatively high temperature fee-standing postcures.

分子量;黏度
构造;曲率
压力;热压罐
后固化

(3) Polyesters.

Polyesters (shown in Fig. 3-6) are low-cost resin systems and offer excellent corrosion resistance. The operating service temperatures for polyesters are lower than for epoxies. In general, for a fiber-reinforced resin system, the advantage of polyester is its low cost and its ability to be processed quickly. Fiber-reinforced polyesters can be processed by many methods. Common processing methods include matched metal molding, wet lay-up, press (vacuum bag) molding, injection molding, filament winding, pultrusion, and autoclaving. Polyesters can be formulated to cure more rapidly than phenolics during the thermoset molding process. **While phenolic processing, for example, is dependent on a time/temperature relationship, polyester processing is primarily dependent on temperature.** Depending on the formulation, polyesters can be processed from room temperature to 350°F (180℃). If the proper temperature is applied, a quick cure will occur. Without sufficient heat, the resin/catalyst system will remain plasticized. Compared to epoxies, polyesters process more easily and are much tougher, whereas phenolics are more difficult to process and brittle, but have higher service temperatures.

耐腐蚀性
纤维增强树脂
纤维增强聚酯
匹配的;手糊;压力(真空袋)成型
注射;缠绕;拉挤;热压罐;配方;
工艺过程;
关系
有效的;催化剂
塑化
韧性的
脆的

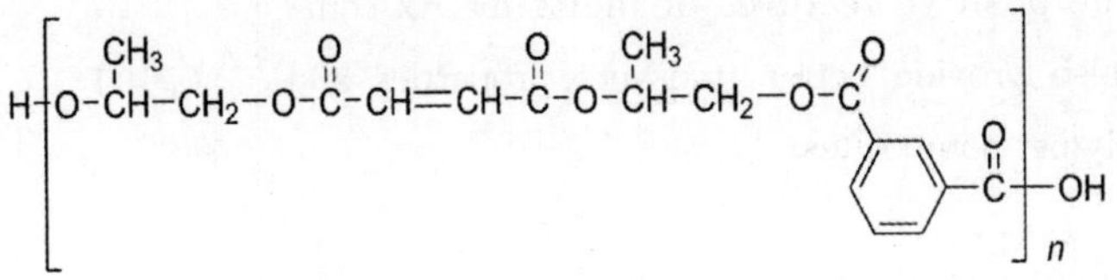

Fig. 3 - 6　The structure of polyesters

(4)　Thermoplastic resins.

Thermoplastic materials are, in general, ductile and tougher than thermoset materials and are used for a wide variety of nonstructural applications without fillers and reinforcements. Thermoplastics can be melted by heating and solidified by cooling, which render them capable of repeated reshaping and reforming. Thermoplastic molecules do not cross-link and therefore they are flexible and reformable. The common thermoplastic resins are showed in Table 3 - 2.

柔软的
非结构的
填料
促使
(聚合物)交联;再成型

Table 3 - 2　Typical unfilled thermoplastic resin properties

Resin Material	**Density** g/cm^3	**Tensile Modulus** GPa (10^6 psi)	**Tensile Strength** MPa (10^3 psi)
Nylon	1.1	1.3 - 3.5(0.2 - 0.5)	55 - 90(8 - 13)
PEEK	1.3 - 1.35	3.5 - 4.4(0.5 - 0.6)	100(14.5)
PPS	1.3 - 1.4	3.4 (0.49)	80(11.6)
Polyester	1.3 - 1.4	2.1 - 2.8(0.3 - 0.4)	55 - 60(8 - 8.7)
Polycarbonate	1.2	2.1 - 3.5(0.3 - 0.5)	55 - 70(8 - 10)
Acetal	1.4	3.5 (0.5)	70(10)
Polyethylene	0.9 - 1.0	0.7 - 1.4(0.1 - 0.2)	20 - 35(2.9 - 5)
Teflon	2.1 - 2.3	—	10 - 35(1.5 - 5.0)

2. Metal Matrix

Polymer composites are used normally up to 180℃, but rarely beyond 350℃. The high temperature capabilities of inorganic reinforcements can not be realized, when polymers are employed as matrix material. **Metal matrices, on the other hand, can widen the scope of using composites over a wide range of temperatures.** Besides, metal matrix composites allow tailoring of several useful properties that are not achievable in conventional metallic alloys. High specific strength and stiffness, low thermal expansion, good thermal stability and improved

无机的
范围
调整

wear resistance are some of the positive features of metal matrix composites. The metal composites also provide better transverse properties and higher toughness compared to polymer composites.

横向的
韧性

3. Ceramic Matrix

Ceramic materials are very hard and brittle. Generally, they consist of one or more metals combined with a nonmetal such as oxygen, carbon, or nitrogen. They have strong covalent and ionic bonds and very few slip systems available compared to metals. Thus, characteristically, ceramics have low failure strains and low toughness or fractureenergies. **In addition to being brittle, they lack uniformity in properties, have low thermal and mechanical shock resistance, and have low tensile strength.** On the other hand, ceramic materials have very high elastic modulus, low densities, and can withstand very high temperatures. The last item is very important and is the real driving force behind the effort to produce tough ceramics. Consider the fact that up to 800℃ and can go up to 1,100℃ with oxidation-resistant coatings. Beyond this temperature, one must use ceramic materials.

非金属
氮；共价键；离子键
失效应力；韧性；断裂；一致性
震荡
弹性模量；承受
抗氧化

A composite is defined as a material containing two or more distinct phases combined in such a way so that each remains distinct. Based on this broad definition of a composite ceramic matrix composite (CMCs) are conveniently separated into two categories: discontinuous reinforced and continuous fiber reinforced CMCs.

截然不同的
定义；复合的
不连续的

Notes

1. The advantages of epoxies are high strength and modulus, low levels of volatiles, excellent adhesion, low shrinkage, good chemical resistance, and ease of processing. Their major disadvantages are brittleness and the reduction of properties in the presence of moisture.

环氧树脂的优点是强度和模量高，挥发物含量低，附着力好，收缩率低，耐化学性好，加工方便。它的主要缺点是脆性和在潮湿环境下性能下降。

2. Epoxy-based composites provide good performance at room and elevated temperatures. Curing temperatures vary from room temperature to approximately 350°F (180℃). The most common cure temperatures range between 250°F and 350°F (120℃ and 180℃). The use temperatures of the cured structure will also vary with the cure temperature.

环氧基复合材料在室温和高温下均具有良好的性能。固化温度从室温至大约 350°F (180℃)。最常用的固化温度区间在 250°F 到 350°F 之间(120℃到 180℃)。固化后制品的使用温度也会随着固化温度的变化而变化。

3. Epoxies come in liquid, solid, and semi-solid forms. Liquid epoxies are used in RTM, filament winding, pultrusion, hand lay-up, and other processes with various reinforcing fibers such as glass, carbon, aramid, boron, etc.

环氧树脂有液体、固体和半固体三种形式。液体环氧树脂与玻璃、碳、芳纶、硼等多种增强纤维一起应用于RTM、缠绕、拉挤、手糊等工艺。

4. The phenolic polymers, often called either phenolic resole or novolacs resins are condensation polymers based upon either a reaction of formaldehyde with a base catalyst and phenol (resole), or a reaction of excess phenol with an acidic catalyst and formaldehyde (novolac).

通常把酚醛聚合物称为热固性酚醛树脂或者热塑性酚醛树脂，它是甲醛和苯酚在碱性催化剂下或过量苯酚和甲醛在酸性催化剂下的反应而形成的缩聚聚合物。

5. The basic difference between resoles and novolacs consist of no methylol groups in the novolacs and the resulting need for an extension agent of paraformaldehyde, hexamethylene-tetramine, or additional formaldehyde as a curative.

热固性酚醛树脂和热塑性酚醛树脂的基本区别在于，热塑性酚醛树脂中不含甲基醇，因此需要多聚甲醛、六亚甲基四胺或过量的甲醛作为一种引发剂。

6. While phenolic processing, for example, is dependent on a time/temperature relationship, polyester processing is primarily dependent on temperature.

例如，虽然酚醛加工取决于时间和温度关系，但聚酯加工主要取决于温度。

7. Metal matrices, on the other hand, can widen the scope of using composites over a wide range of temperatures.

另一方面，金属基体可以扩大复合材料使用的温度范围。

8. In addition to being brittle, they lack uniformity in properties, have low thermal and mechanical shock resistance, and have low tensile strength.

除易碎外，它们的性能缺乏均匀性，耐热性、机械冲击性以及拉伸强度较低。

Exercises

1. Translate the following sentences into Chinese.

(1)It is important to have a good knowledge of the various raw materials available for the manufacture of good composite products.

(2)In general, raw materials for composite manufacturing processes can be divided into two categories: thermoset-based and thermoplastic-based composite materials.

(3)Matrix surrounds the fibers and thus protects those fibers against environmental and chemical attack.

(4)Because of predominantly covalent bounding, polymers are generally poor conduction of heat and electricity.

(5)Thermoplastics comprise essentially linear or lightly branched polymer molecules, while thermosets are substantially crosslinked materials, consisting of an extensive three-dimensional network of covalent chemical bonding.

(6)Epoxies are more costly than polyester and vinylesters and are therefore not used in cost-sensitive markets (e. g. automotive and marine) unless specific performance is needed.

(7)Compared to epoxies, polyesters process more easily and are much tougher, whereas phenolics are more difficult to process and brittle, but have higher service temperatures.

(8)Thermoplastic materials are, in general, ductile and tougher than thermoset materials and are used for a wide variety of nonstructural applications without fillers and reinforcements.

(9)Polymer composites are used normally up to 180℃, but rarely beyond 350℃.

(10) They have strong covalent and ionic bonds and very few slip systems available compared to metals.

2. Translate the following into English.

复合材料产品

热塑性树脂

聚合物基复合材料

金属基复合材料

陶瓷基复合材料

碳-碳基复合材料

比强度

比刚度

热膨胀

热稳定

耐磨性

3. Reading comprehension.

(1)Which one is right in the following sentences?

A. Polymers have higher strength and modulus and lower temperature use limits.

B. Polymers have lower strength and modulus and higher temperature use limits.

C. Polymers are generally good conduction of heat and electricity.

D. Polymers are generally poor conduction of heat and electricity.

(2)Which of the following is the characteristic of thermoplastic?

A. They can not be repeatedly melted or reprocessed.

B. They can be repeatedly melted or reprocessed.

C. They are characterized by branched chain molecules.

D. Thermal exposure won't degrade their properties.

(3)Which of the following statements is right according to the passage?

A. The processing or curing of epoxies is slower than polyester resins.

B. An epoxide group is an oxirane structure, a three-member ring with two oxygen and one carbon atoms.

C. The advantages of epoxies are low strength and modulus, slow levels of volatiles, excellent adhesion, high shrinkage, good chemical resistance, and ease of processing.

D. Epoxies are polymerizable thermoplastic resins.

(4) All of the following methods are common processing methods of fiber-reinforced polyesters except ________.

A. matched metal molding

B. press (vacuum bag) molding

C. injection molding

D. extrusion molding

(5) The phenolic polymers is ________.

A. a less expensive alternative for lower temperature use

B. condensation polymers based upon either a reaction of excess formaldehyde with a base catalyst and phenol(resole), or a reaction of excess phenol with an acidic catalyst and formaldehyde(novalac)

C. characterized by both chemical and thermal resistance as well as hardness, and low smoke and toxic degradation products.

D. dependent on a time/temperature relationship

(6) Which of the following resins is processed by the reaction of phenol and excess formaldehyde in the presence of base?

A. Resoles B. Novolacs C. Bismaleimide D. Epoxies

Reading Material

There are two major groups of resins that make up what we call polymer materials-**thermosets** and **thermoplastics.** These resins are made of polymers (large molecules made up of long chains of smaller molecules or monomers).

Thermoset resins are used to make most composites. They're converted from a liquid to a solid through a process called polymerization, or cross-linking. When used to produce finished goods, thermosetting resins are "cured" by the use of a catalyst, heat or a combination of the two. Once cured, solid thermoset resins cannot be converted back to their original liquid form. Common thermosets are polyester, vinyl ester, epoxy, and polyurethane (as shown in Fig. 3-7).

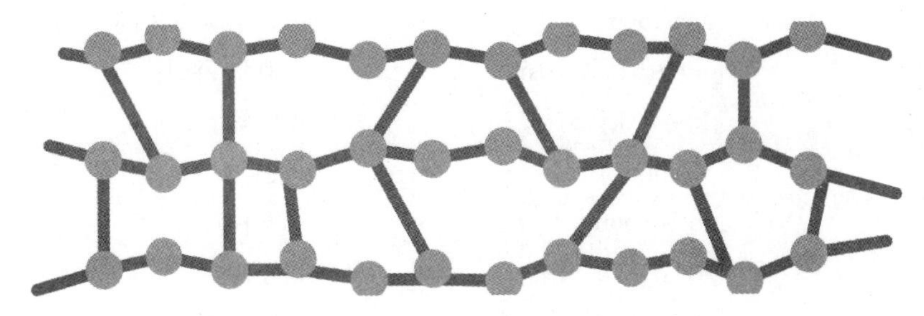

Fig. 3-7 Thermosets cross link during the curing process to form an irreversible bond

Thermoplastic resins, on the other hand, are not cross-linked and, so, can be melted, formed, re-melted and re-formed. Thermoplastic resins are characterized by materials such as ABS, polyethylene, polystyrene, and polycarbonate (as shown in Fig. 3-8).

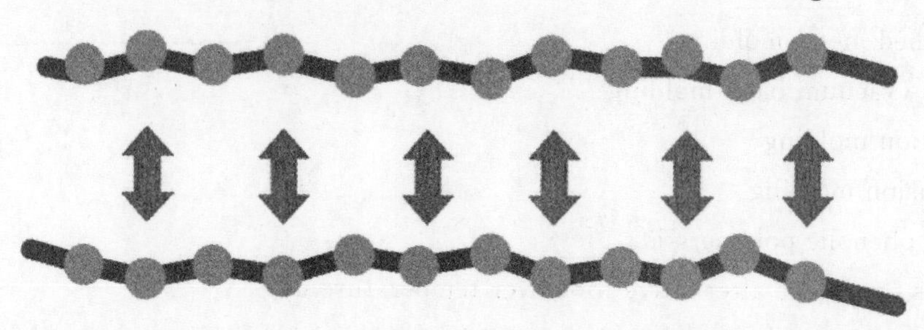

Fig. 3-8　Thermoplastics form extremely strong bonds within chain molecules

These resins are recognized by their capability to be shaped or molded while in a heated semi-fluid state and become rigid when cooled. We are surrounded by everyday household items made of thermoplastics.

New Words and Expressions

raw	[rɔː]	adj. 未加工的
corresponding	[kɒrɪˈspɒndɪŋ]	adj. 相当的,相应的
solidified	[səˈlidəfaid]	adj. 凝固的;固化的,变硬的
storage	[ˈstɔːrɪdʒ]	n. 存储
recyclability	[riːsɪkləˈbɪlɪtɪ]	n. 再循环能力
ingredients	[ɪnˈgriːdɪənts]	n. 配料;材料
matrix	[ˈmeɪtrɪks]	n. 基体材料
surround	[səˈraʊnd]	v. 包裹
attack	[əˈtæk]	v. 攻击,侵蚀
modulus	[ˈmɒdjʊləs]	n. 模量
elongation	[iːlɒŋˈgerʃ(ə)n]	n. 延伸
relevant	[ˈreləvənt]	adj. 相关的
processed	[ˈprəʊsest]	adj. 已加工过的;加工的
prolonged	[prəˈlɒŋd]	adj. 持续很久的
degradation	[ˌdegrəˈdeɪʃ(ə)n]	n. 毁坏;恶化
conduction	[kənˈdʌkʃ(ə)n]	n. 传导;输送
giant	[ˈdʒaɪənt]	adj. 巨大的
backbone	[ˈbækbəʊn]	n. 支柱
treatment	[ˈtriːtm(ə)nt]	n. 处理
decompose	[diːkəmˈpəʊz]	v. 分解

distinction	[dɪ'stɪŋ(k)ʃ(ə)n]	n. 差别；对比
comprise	[kəm'praɪz]	v. 包括
crosslinked	[k'rɒslɪŋkt]	adj. 交联的
polymerization	[ˌpɒlɪmərai'zeɪʃn]	n. 聚合
condensation	[kɒnden'seɪʃ(ə)n]	n. 浓缩；缩聚
addition	[ə'dɪʃ(ə)n]	n. 加成
monomers	['mɒnəʊməz]	n. 单体
loss	[lɒs]	n. 损失
unsaturated	[ʌn'sætʃəˌreɪtɪd]	adj. 不饱和的
generic	[dʒɪ'nerɪk]	adj. 一般的；普通的
designate	['dezɪgneɪt]	v. 指定；标出
versatile	['vɜːsətaɪl]	adj. 通用的；万能的
shrinkage	['ʃrɪŋkɪdʒ]	n. 收缩率
adhesion	[əd'hiːʒ(ə)n]	n. 黏附；支持；固定
substrate	['sʌbstreɪt]	n. 基底；基层
adhesives	[əd'hiːsivz]	n. 黏合剂
volatiles	['vəlætəliːz]	n. 挥发组分
brittleness	['brɪtlnɪs]	n. 脆性，脆度
moisture	['mɒɪstʃə]	n. 湿度；潮湿
polyester	[ˌpɒlɪ'estə]	n. 聚酯
formulate	['fɔːmjuleɪt]	v. 系统地阐述
specific	[spə'sɪfɪk]	adj. 特殊的
elevated	['elɪveɪtɪd]	adj. 升高的；高层的
cure	[kjʊr]	v. 固化
yield	[jiːld]	v. 生产
aramid	['ærəmld]	n. 芳香族聚酰胺
boron	['bɔːrɒn]	n. [化]硼
capsules	['kæpsl]	n. 胶囊、颗粒
toxic	['tɒksɪk]	adj. 有毒的
formaldehyde	[fɔː'mældɪhaɪd]	n. 甲醛
base	[beɪs]	adj. 碱性的
phenol	['fiːnɒl]	n. 酚
acidic	[ə'sɪdɪk]	adj. 酸的，酸性的
paraformaldehyde	[ˌpærəfɔr'mældəhaɪd]	n. 多聚甲醛
hexamethylenetetramine	['heksəmeθəliːn'tetrəmiːn]	n. 六亚甲基四胺

viscosity	[visˈkɔsiti]	n. 黏度
conformation	[kɒnfɔːˈmeɪʃ(ə)n]	n. 构造
curvature	[ˈkɜːvətʃə]	n. 弯曲，[数]曲率
tailoring	[ˈteɪlərɪŋ]	v. 剪裁
withstand	[wɪðˈstænd]	v. 承受；抵住；顶住；经受住
distinct	[dɪˈstɪŋkt]	adj. 明显的；截然不同的
methylol		羟甲基；亚甲醇
cost-sensitive		成本敏感
flame-resistant		耐火的；抗火的
corrosion resistance		耐腐蚀性
fiber-reinforced resin		纤维增强树脂
composite manufacturing techniques		复合材料制造技术
thermoset-based		热固性基的
thermoplastic-based		热塑性基的
reinforcing fibers		增强纤维
covalent bounding		共价键
phenolic resole		酚醛树脂
vinyl chloride		氯乙烯
oxidation-resistant		抗氧化

Unit 2 Reinforcements

Reinforcements are important constituents of a composite material and give all the necessary stiffness and strength to the composite. These are thin rodlike structures. The most common reinforcements are glass, carbon, aramid and boron fibers. (as shown in Fig. 3 - 9). Typical fiber diameters range from 5 μm to 20 μm. The diameter of a glass fiber is in the range of 5 μ to 25 μm, a carbon fiber is 5 μm to 8 μm, an aramid fiber is 12.5 μm, and a boron fiber is 100μm. Because of this thin diameter, the fiber is flexible and easily conforms to various shapes. **In composites, the strength and stiffness are provided by the fibers. The matrix gives rigidity to the structure and transfers the load to fibers.** The reinforcements carry the load. In a structural composite, 70% to 90% of the load is carried by fibers. The properties of fibers and conventional bulk materials are shown in Table 3 - 3.

增强体；成分
硬度
棒状
硼纤维
直径范围
柔软的；顺从
刚度
载荷
块材

Table 3 - 3 Properties of fibers and conventional bulk materials

Material		Diameter / μm	Density (ρ) / g/cm^3	Tensile Modulus (E) / GPa	Tensile Strength (σ) / GPa	Specific Modulus (E/ρ) / (m^2/s^2)	Specific Strength (σ/ρ) / (m^2/s^2)	Melting Point / ℃	Elongation break (%)	Relative Cost
Fibers	E-glass	7	2.54	70	3.45	27	1.35	>1,540	4.8	Low
	S-glass	15	2.50	86	4.50	34.5	1.8	>1,540	5.7	Moderate
	Graphite, high modulus	7.5	1.9	400	1.8	200	0.9	>3,500	1.5	High
	Graphite, high strength	7.5	1.7	240	2.6	140	1.5	>3,500	0.8	High
	Boron	130	2.6	400	3.5	155	1.3	2,300	—	High
	Kevlar 29	12	1.45	80	2.8	55.5	1.9	500(D)	3.5	Moderate
	Kevlar 49	12	1.45	130	2.8	89.5	1.9	500(D)	2.5	Moderate
Bulk materials	Steel	—	7.8	208	0.34 - 2.1	27	0.0 - 0.27	1,480	5 - 25	Low
	Aluminum	—	2.7	69	0.14 - 0.62	26	0.05 - 0.23	600	8 - 16	Low

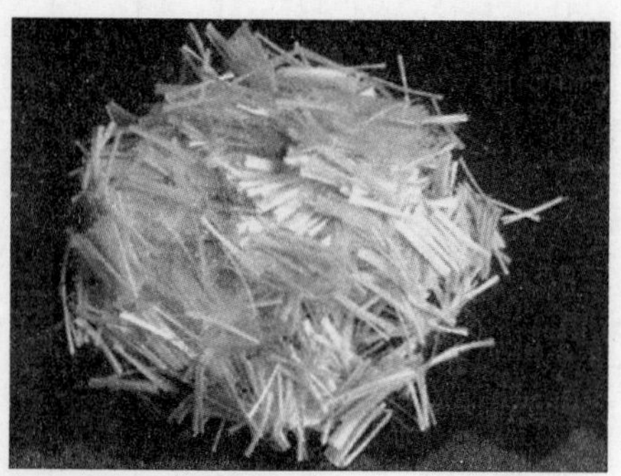

Fig. 3 - 9 The most common reinforcements

1. Glass Fiber

(1) What is glass fibers?

Glass fiber (formerly known as: glass fiber or fiberglass) is an inorganic non-metallic material with excellent performance. **It has a wide variety of advantages, such as good insulation, high heat resistance, good corrosion resistance and high mechanical strength, but its disadvantages are brittleness and poor wear resistance. It is a glass ball or waste glass as raw material by high temperature melting, drawing, winding, weaving and other processes into the manufacture, the diameter of its single filament for a few microns to 20 microns, equivalent to a hair 1/20 - 1/5, each bundle of filaments are composed of hundreds or even thousands of single filament.** Glass fiber is commonly used as reinforced materials in composite materials, electrical insulation materials and thermal insulation materials, and other areas of the national economy.

无机的;非金属的;性能;绝缘;耐热
抗腐蚀性
脆性;耐磨性
牵引;缠绕;编织
微米;等效的
单丝
电的;热的

(2) What are the main features of glass fiber?

Raw materials and applications: glass fiber have high temperature resistance than organic fiber, non-combustible, anti-corrosion, heat insulation, sound insulation, high tensile strength, good electrical insulation. But they are brittle, poor wear resistance. Used to make reinforced plastics or reinforced rubber as reinforcing material, glass fiber has the following characteristics. These characteristics make the

不易燃的
橡胶
特征

use of glass fiber far more than other types of fiber, and the development speed is far ahead of. Its characteristics are listed as follows:

①High tensile strength, small elongation(3%).

抗张强度;伸长率

②High elastic coefficient, good rigidity.

弹性系数

③ Large elongation within the elastic limit and high tensile strength, so absorb impact energy.

冲击能

④It is an inorganic fiber with non-flammability and good chemical resistance.

不燃性

⑤Low water absorption.

吸收

⑥Good scale stability and heat resistance.

稳定性

⑦Good process ability, can be made into strands, bundles, blankets, weaving and other different forms of products.

线;束;毛毯
纺织物

⑧Transparent through the light.

透明的

⑨Development of a surface treatment agent with good adhesion to the resin has been completed.

处理剂

⑩The price is cheap.

⑪Not easy to burn, high temperature can be fused into glass beads.

玻璃珠

(3)What are the types of glass fibers?

Glass fibers are divided into different grades according to their composition, properties and uses. According to the standard grade, class E glass fiber is the most commonly used, widely used in electrical insulation materials; S grade is a special fiber, although the output is small, but very important, because of its super strength, mainly used for military defense, such as bulletproof box; Class C is more chemical resistant than class E, used for battery isolation plate, chemical filter; Grade A is basic glass fiber, used to produce reinforced materials.

合成物
产量
军事防御;防弹盒子
电池绝缘;过滤器

The glass used to make fiberglass is different from other glass products. The composition of glass for internationally commercialized fibers is as follows:

E-glass, also known as non-alkali glass, it is a boron silicate glass. At present, it is the most widely used glass component for glass fiber, with good electrical insulation and mechanical properties. It is widely used in the production of glass fiber for electrical insulation and glass fiber reinforced plastics.

无碱;硼硅酸盐

C-glass, also known as alkali glass, its characteristics are chemical resistance, especially acid resistance is better than non-alkali glass, but poor electrical performance, mechanical strength is lower than non-alkali glass fiber 10%-20%, usually the foreign alkali glass fiber

耐酸性

containing a certain amount of boron oxide and Chinese alkali glass fiber is completely boron free. In foreign countries, alkali glass fiber is only used to produce corrosion-resistant glass fiber products, such as glass fiber surface felt, etc.

二氧化硼

毡制品

But in China, alkali glass fiber accounts for more than half (60%) of the output of glass fiber. It is widely used in the reinforcement of glass fiber reinforced plastics and the production of filter fabric and wrapping fabric, etc. , because its price is lower than that of non-alkali glass fiber, it has strong competitiveness.

竞争力

(4) What are glass fiber-reinforced composites?

The technical demand for engineering polymers with higher temperature resistance, reasonable toughness and ease of processing has led to the development of several high-performance thermoplastics often reinforced with glass fibers and other reinforcement. **By incorporation a small amount of glass fibers, in the range of 10 to 30 weight percent, the mechanical properties, together with corrosion resistance and electrical properties can be improved dramatically.** The properties of the final product depend on concentration of fibers incorporated and on the properties of the fibers including fiber length and ability to transfer stress across the fiber length.

热塑性塑料

浓度

Glass is popular as a fiber reinforcement material for several reasons:

① It is easily drawn into high-strength fibers from the molten state.

熔化的

②It is readily available and may be fabricated into a glass-rein relatively strong, and when embedded in a plastic matrix, it produces a composite having a very high specific strength.

可获得的;制造

嵌入式的

比强度

③When coupled with the various plastics, it possesses a chemical inertness that renders the composite useful in a variety of corrosive environments.

惰性;腐蚀的

2. Carbon Fiber

(1) What is Carbon fiber?

Carbon is a high-performance fiber material that is themost commonly used reinforcement in advanced polymer-matrix composites. Fiber diameters normally range between 4 - 6um. **The carbon fibers used are required to possess high stiffness, high modulus, elasticity, high fatigue strength and low thermal expansion coefficient.** Based on starting materials, carbon fibers are classified as polyacrylonitrile (PAN)-based and pitch-based fibers.

弹性

疲劳;膨胀系数

(2)What are the properties of carbon fiber?

Carbon fiber has the characteristics of strong tensile strength of carbon material and softness and process ability of fiber. The carbon fiber is a new material with excellent mechanical properties. The tensile strength of carbon fiber is about 2 - 7GPa, and the tensile modulus is about 200 - 700GPa. The density is about 1.5 - 2.0 g/cm^3, which depends on the temperature of carbon conduct addition to the structure of the filament. Generally after high temperature 3,000 ℃ treatment, the density of 2.0 g/cm^3. Combined with its light weight, its specific gravity is lighter than aluminum and less than a quarter of that of steel, which gives carbon fiber the highest specific strength and modulus of any high-performance fiber. Compared with titanium, steel, aluminum and other metal materials, carbon fiber has the characteristics of high strength, high modulus, low density and small linear expansion coefficient in physical properties, which can be called the king of new materials.

柔软;成型

拉伸模量

比强度

钛

线型的

膨胀系数

In addition to the general properties of carbon materials, the shape is significantly soft, can be processed into a variety of fabrics, and due to the small specific gravity, along the direction of the fiber axis shows a high strength, carbon fiber reinforced epoxy resin composite material, its specific strength, specific modulus comprehensive index, is the highest in the existing structure material.

纤维织物

比重

综合指数

(3)Where are the main applications of carbon fiber?

1)Composite materials.

Carbon fiber is used as thermal insulation material in addition to the traditional use. It is mainly used as reinforcing material to add resins, metals, ceramics, concrete and other materials to form composite materials. Carbon fiber has become the most important reinforcement of advanced composite materials.

绝缘

2)Civil engineering and construction.

In the field of civil engineering and architecture, carbon fiber is also used to reinforce industrial and civil buildings, railway, highway, bridge, tunnel, chimney, tower structure, etc. In the railway construction, large roof system and sound insulation wall will have good applications in the future, which will also be a very promising application of carbon fiber.

隧道;烟囱

3)Aeronautics and astronautics.

Carbon fiber is the strategic base material for rocket, satellite, missile, fighter and ship.

战略的

导弹

4) Auto materials.

In car interior and exterior adornment begin to use in great quantities. Carbon fiber used as a car material, the biggest advantage is light weight, strength, weight is only equivalent to (20 - 30)% percent of steel, steel hardness is more than 10 times.

装饰 相等的 硬度

5) Sports goods.

Carbon fiber is used in the field of sports and leisure, such as golf clubs, fishing rods, tennis rackets, badminton rackets, bicycles, ski poles, skis, sailboards, masts, sailing hulls and other sporting goods are one of the main users of carbon fiber.

闲暇 帆船

The carbon fiber in addition to aerospace, defense and military applications and sporting goods, carbon fiber is being used in new markets such as automotive components, wind power blades, building reinforcement materials, reinforced plastics and drilling platforms. In addition, it is also used in pressure vessels, medical equipment, Marine development, new energy and other fields. Other applications of carbon fiber include the production of composite materials for machine parts, household appliances and semiconductor related equipment, which can be used to enhance protection against static electricity and electromagnetic waves.

软管 半导体 静电 电磁的

3. Aramid fiber

(1) What is aramid fiber?

Aramid fiber and Kevlar belong to a family of synthetic products characterized by strength. It is appropriate for various applications such as composites, bullet-proof products, building materials, special protective clothing, electronic equipment and other fields.

芳纶;凯芙拉;合成的 防弹的

(2) What are the main features of aramid fiber?

Aramid fibers are characterized by excellent environmental and thermal stability, static and dynamic fatigue resistance, and impact resistance. These fibers have the highest specific tensile strength (strength/density ratio) of any commercially available continuous-filament yarn. Aramid-reinforced thermoplastic composites have excellent wear resistance and near-isotropic properties—characteristics not available with glass or carbon-reinforced composites.

环境的 热的;静态的;动态 疲劳;冲击; 比强度 长丝纱线 近均相性

(3) What is the history of Aramid Fibers?

Aramid fibers, trade name Kevlar (Du Pont), were the first organic fibers with sufficient stiffness and strength for use in PMCs suitable for airframe applications. Kevlar aramid fibers significantly exceed the specific strength and stiffness of glass fibers. PMCs based

有机的;足够的 超过

on aramid fibers have attractive tensile properties for temperatures of over 400℃; however, they have poor compression strength. This is a major limitation for applications subject to significant compression loading, including those requiring high bearing strength.

吸引人的
抗压强度
限制;压缩载荷
承载强度

A major advantage of aramid fibers is their ability to absorb amounts of energy during fracture, which results from their high strain-to-failure, their ability to undergo plastic deformation in compression, and their ability to defibrillate during tensile fracture. The fibers exhibit plastic behavior at around 0.3% compression strain, and deformation is linear up to failure at strains greater than 2% in tension. Hence aramid PMCs are used for ballistic protection and also for engine containment rings. The most common use of aramid PMCs in aerospace has been in fairings, but it is also used as the skins or face sheets for honeycomb panels. Aramid PMCs are also used in radomes and other applications requiring structural efficiency and suitable dielectric properties (as shown in Fig. 3-10 and Fig. 3-11).

破裂
应变失效;经历;变形
除颤
射击的
环
整流罩
蜂窝;面板;雷达罩
功效

Fig. 3-10 Aramid PMCs used for ballistic protection

Fig. 3-11 Aramid honeycomb

Notes

1. Reinforcements are important constituents of a composite material and give all the necessary stiffness and strength to the composite.

增强材料是复合材料的重要组成部分,为复合材料提供所需要的刚度和强度。

2. In composites, the strength and stiffness are provided by the fibers. The matrix gives rigidity to the structure and transfers the load to fibers.

复合材料的强度和硬度由纤维提供。基体决定了结构刚度，并将载荷传递给纤维。

3. It has a wide variety of advantages, such as good insulation, high heat resistance, good corrosion resistance and high mechanical strength, but its disadvantages are brittleness and poor wear resistance.

它具有很多优点，比如绝缘性好、耐热性高、耐腐蚀性好、机械强度高，但缺点是脆性大、耐磨性差。

4. It is a glass ball or waste glass as raw material by high temperature melting, drawing, winding, weaving and other processes into the manufacture, the diameter of its single filament for a few microns to 20 μm, equivalent to 1/20 - 1/5 of a hair , each bundle of filaments are composed of hundreds or even thousands of single filament.

它是以玻璃球或废玻璃为原料，经高温熔融、拉拔、缠绕、编织等工艺制成，其单丝直径为几微米至 20μm，相当于一根头发的 1/20～1/5，每束细丝由数百根或甚至几千根单丝组成。

5. By incorporation a small amount of glass fibers, in the range of 10 - 30 weight percent, the mechanical properties, together with corrosion resistance and electrical properties can be improved dramatically.

通过掺入少量的玻璃纤维，其质量分数在 10％到 30％时，复合材料机械性能、耐腐蚀性和电性能都能得到显著改善。

6. The carbon fibers used are required to possess high stiffness, high modulus, elasticity, high fatigue strength and low thermal expansion coefficient.

所用的碳纤维要求具有高刚度、高模量、高弹性、高疲劳强度和低热膨胀系数。

Exercises

1. Translate the following sentences into Chinese.

(1) Glass fibers are divided into different grades according to their composition, properties and uses.

(2) It is widely used in the production of glass fiber for electrical insulation and glass fiber reinforced plastics.

(3) The properties of the final product depend on concentration of fibers incorporated and on the properties of the fibers including fiber length and ability to transfer stress across the fiber length.

(4) The properties of the final product depend on concentration of fibers incorporated and on the properties of the fibers including fiber length and ability to transfer stress across the fiberlength .

(5) Carbon is a high-performance fiber material that is themost commonly used reinforcement in advanced polymer-matrix composites.

(6) Combined with its light weight, its specific gravity is lighter than aluminum and less than a quarter of that of steel, which gives carbon fiber the highest specific strength and

modulus of any high-performance fiber.

(7)It is appropriate for various applications such as composites, bullet-proof products, building materials, special protective clothing, electronic equipment and other fields.

(8)A major advantage of aramid fibers is their ability to absorb amounts of energy during fracture, which results from their high strain-to-failure, their ability to undergo plastic deformation in compression, and their ability to defibrillate during tensile fracture.

2. Translate the following into English.

绝缘性好

耐热性高

耐腐蚀性好

机械强度高

疲劳强度

热膨胀系数

塑性变形

蜂窝面板

3. Reading comprehension.

(1)What do the properties of the final product depend on?

A. the concentration of fibers incorporated

B. the properties of the fibers

C. ability to transfer stress across the fiber length

D. A, B and C

(2)Which one is not the advantages of glass fibers?

A. high heat resistance.

B. good corrosion resistance.

C. good wear resistance.

D. high mechanical strength.

(3)What is not the type of glass fibers?

A. E-glass B. A-glass C. S-glass D. C-glass

(4)What is the Prominent characteristics of C-glass?

A. good acid resistance.

B. good electrical performance.

C. high mechanical strength.

D. low cost.

(5)The strength of the carbon-carbon composites is mainly dependent on ________.

A. architecture of fibers in the composites

B. tensile strength of fiber

C. the content of fiber

D. the strength of thematrice

(6) What does make carbon-carbon composites quite attractive in certain high-performance area?

A. low thermal expansion coefficient

B. good mechanical integrity

C. high heat storage capacity

D. good chemical resistance

(7) Which is not correct about the selection of carbon fibers for carbon-carbon composites?

A. High modulus fibers lead to carbon-carbon composites with better mechanical properties and good oxidation resistance.

B. The difference in fiber structures significantly influences the properties of the fibers and leads to different fiber/matrix interactions.

C. High strength carbon fibers do necessarily produce carbon-carbon composites.

D. Pitch-based fibers have considerable potential as reinforcement of carbon-carbon composites.

(8) Which is not the main applications of Carbon Fiber?

A. automotive components　　　B. wind power blades

C. pressure vessels　　　D. A, B and C

(9) What is a major advantage of aramid fibers?

A. excellent environmental and thermal stability

B. static and dynamic fatigue resistance

C. their ability to absorb amounts of energy during fracture

D. the highest specific tensile strength

(10) Which are the main applications of aramid fiber?

A. composites

B. bullet-proof products

C. building materials

D. special protective clothing, electronic equipment and other fields

Reading Material

Prepregs

A prepreg is a resin-impregnated fiber, fabric, or mat in fiat form, which is stored for later use in hand lay-up or molding operations. Fibers laid at horizontal direction and pre-impregnated with resin are called unidirectional tape. It shows the various types of prepregs available as unidirectional tape, woven fabric tape and roving (as shown in Fig. 3-12).

Epoxy-based prepregs are very common in industry and come in flat sheet form in a thickness range of 0.127-0.254 mm. Prepregs can be broadly classified as thermoset-based prepregs and thermoplastic-based prepregs/tapes. The difference between the two prepregs is the type of resin used. Reinforcements in a prepreg can be glass, carbon or aramid, are used in filament or woven fabric or mat form in either type of prepreg.

Fig. 3 - 12 Prepreg types: unidirectional tape, woven fabric prepregs and roving

Prepregs are used in a wide variety of applications, including aerospaceparts, sporting goods, printed circuit boards, medical components, and industrial products. The advantages of prepreg materials over metals are their higher specific stiffness, specific strength, corrosion resistance, and faster manufacturing. The major disadvantage of prepreg materials is their higher cost. Products made with prepreg materials provide a higher fiber volume fraction than those made by filament winding and pultrusion. Prepregs also provide more controlled properties and higher stiffness and strength properties than other composite products.

New Words and Expressions

dielectric	[ˌdaɪɪ'lektrɪk]	adj. 非传导性的;诱电性的
reinforcements	[ˌriːˌin'fɔːˌsmənts]	n. 增强体;加固物
constituents	[kən'stɪtʃʊənt]	n. 成分
stiffness	['stɪfnɪs]	n. 僵硬;坚硬;不自然;顽固
flexible	['flɛksəbl]	adj. 灵活的;柔韧的;易弯曲的
rigidity	[rɪ'dʒɪdəti]	n. [物]硬度,刚性
load	[lod]	n. 负载,负荷
bulk	[bʌlk]	n. 体积,容量;大多数,大部分
inorganic	['ɪnɔr'gænɪk]	adj. 无机的;无生物的
performance	[pə'fɔrməns]	n. 性能;绩效;表演;执行;表现
insulation	[ˌɪnsə'leiʃən]	n. 绝缘;隔离,孤立
brittleness	['brɪtlnɪs]	n. 脆性,脆度;脆弱性
drawing	['drɔɪŋ]	v. 牵引;拖曳

winding	[ˈwaɪndɪŋ]	v. 缠绕;卷绕;被弯曲
equivalent	[ɪˈkwɪvələnt]	adj. 相等的;等价的;等效的
filaments	[ˈfɪləmənt]	n. 粗纱线
electrical	[ɪˈlɛktrɪkl]	adj. 有关电的;电气科学的
rubber	[ˈrʌbə]	n. 橡胶;合成橡胶
characteristics	[ˌkærəktəˈrɪstɪks]	n. 特性,特征;特色;特质
elongation	[ɪˌlɔŋˈgeʃən]	n. 伸长;伸长率;延伸率;延长
absorption	[əbˈsɔrpʃən]	n. 吸收
transparent	[trænsˈpærənt]	adj. 透明的;显然的
treatment	[ˈtritmənt]	n. 治疗,疗法;处理;对待
agent	[ˈedʒənt]	n. 试剂,药剂
beads	[biːdz]	n. 玻璃珠
composition	[ˌkampəˈzɪʃən]	n. 构成;合成物;成分
output	[ˈaʊtpʊt]	n. 输出,输出量;产量;出产
battery	[ˈbæt(ə)rɪ]	n. 电池,蓄电池
isolation	[ˌaɪsəˈleʃən]	n. 隔离;孤立;绝缘;离析
borosilicate	[ˌbɔroˈsɪlɪkɪt]	n. 硼硅酸盐
competitiveness	[kəmˈpɛtətɪvnɪs]	n. 竞争力,好竞争
concentration	[ˈkɑnsnˈtreʃən]	n. 浓度
molten	[ˈmoltən]	adj. 熔化的;铸造的;炽热的
available	[əˈveləbl]	adj. 可获得的;可找到的
fabricate	[ˈfæbrɪˌkeit]	v. 编造;制造
embedded	[ɪmˈbɛdɪd]	adj. 嵌入式的;植入的;内含的
inertness	[iˈnəːtnis]	n. 惰性;无生命力;不活泼
corrosive	[kəˈrosɪv]	adj. 腐蚀的;侵蚀性的
elasticity	[ˌilæˈstɪsəti]	n. 弹性;弹力;灵活性
fatigue	[fəˈtig]	n. 疲劳
softness	[ˈsɔftnɪs]	n. 柔软
linear	[ˈlɪnɪə]	adj. 线的,线型的;直线的
fabrics	[ˈfæbrɪk]	n. 纤维织物
tunnel	[ˈtʌnl]	n. 隧道
chimney	[ˈtʃɪmni]	n. 烟囱
strategic	[strəˈtidʒɪk]	adj. 战略上的,战略的
missile	[ˈmɪsl]	n. 导弹;投射物
adornment	[əˈdɔrnmənt]	n. 装饰;装饰品

hardness	[ˈhardnɪs]	n. 硬度;坚硬
leisure	[ˈliʒə]	n. 闲暇;空闲;安逸
vessels	[ˈvɛslz]	n. 船舶;容器
semiconductor	[ˌsɛmɪkənˈdʌktə]	n. 半导体
electromagnetic	[ɪˌlɛktromægˈnɛtɪk]	adj. 电磁的
synthetic	[sɪnˈθɛtɪk]	adj. 合成的
environmental	[ɪnˌvaɪrənˈmɛntl]	adj. 环境的,周围的
thermal	[ˈθərməl]	adj. 热的
static	[ˈstætɪk]	adj. 静态的
impact	[ɪmˈpækt]	n. 碰撞;冲击力
organic	[ɔrˈgænɪk]	adj. 有机的
sufficient	[səˈfɪʃənt]	adj. 足够的;充分的
exceed	[ɪkˈsid]	vt. 超过;胜过
attractive	[əˈtræktɪv]	adj. 吸引人的;引人注目的
limitation	[ˌlɪmɪˈteʃən]	n. 限制;限度;缺陷
compression	[kəmˈprɛʃən]	n. 压缩
fracture	[ˈfræktʃə]	n. 破裂,断裂
undergo	[ˌʌndərˈgoʊ]	v. 经历
deformation	[diːfɔːrˈmeɪʃn]	n. 变形
defibrillate	[diˈfɪbrɪˈletə̩]	vt. 除颤
ballistic	[bəˈlɪstɪk]	adj. 弹道的;射击的
fairings	[feərɪŋ]	n. 整流罩
honeycomb	[ˈhʌnɪkom]	n. 蜂窝
panels	[ˈpænl]	n. 面板
efficiency	[ɪˈfɪʃənsɪ]	n. 效率;效能
insulation	[ˌɪnsəˈleʃən]	n. 绝缘;隔离,孤立
bullet-proof	[ˈbulitˌpruːf]	adj. 防弹的
aramid	[ˈærəmɪd]	n. 芳族聚酰胺
boron fiber		硼纤维
conform to		遵循,服从
non-metallic		非金属的
heat resistance		耐热性能
corrosion resistance		耐蚀性;抗腐蚀性
wear resistance		耐磨性,耐磨度
strain-to-failure		应变失效

non-combustible	不易燃的
tensile strength	抗张强度
elastic coefficient	弹性系数;弹性常数;弹性模量
impact energy	冲击能
non-flammability	不燃性,耐燃性
military defense	军事防御
bulletroof box	防弹的盒子
non-alkali	无碱
acid resistance	耐酸性
boron oxide	氧化硼;二氧化硼
specific strength	比强度;强度系数
expansion coefficient	膨胀系数,膨胀率;展开系数
tensile modulus	拉伸模量;定伸强度;抗张模量
specific strength	比强度;强度系数
expansion coefficient	膨胀系数,膨胀率;展开系数
specific gravity	比重
comprehensive index	综合指数
dynamic fatigue	动态疲劳
specific tensile strength	比强度
filament yarn	长丝纱线
compression strength	抗压强度;压缩强度
bearing strength	承载强度;承压强度;支承强度

Chapter 4 Manufacturing Processing of Composite Materials

Warm-up Discussion:

①Are you familiar with the common composite processing methods?

②Do you know how to describe the basic steps, the advantages and disadvantages of every process?

③Do you have a good understanding of the difference about the similar process?

Composite materials have succeeded remarkably in their relatively short history. But for continued growth, especially in structural uses, certain obstacles must be overcome. A major one is the tendency of designers to rely on traditional materials such as steel and aluminum unless composites can be produced at lower cost.

障碍;倾向
铝

Cost concerns have led to several changes in the composites industry. There is a general movement toward the use of less expensive fibers. For example, graphite and aramid fibers have largely supplanted the more costly boron in advanced-fiber composites. As important as savings on materials may be, the real key to cutting composite costs lies in the area of processing.

关注
石墨;芳族聚酰胺;代替
硼
在于……

It is a monumental challenge for design and manufacturing engineers to select the right manufacturing process for the production of a part, the reason being that design and manufacturing engineers have so many choices in terms of raw materials and processing techniques to fabricate the part. This section briefly discusses the criteria for selecting a process. Selection of a process depends on the application need. The criteria for selecting a process depend on the production rate, cost, strength, and size and shape requirements of the part, as described in Table 4-1.

非常大的
依据
标准
应用
生产速度
成本

Table 4 - 1 Manufacturing process selection criteria

Process	Speed	Cost	Strength	Size	Shape	Raw Material
Filament winding	Slow to fast	Low to high	High	Small to large	Cylindrical and axisymmetric	Continuous fibers with epoxy and polyester Resins
Pultrusion	Fast	Low to Medium	High(along longitudinal direction)	No restriction on length; small to medium size; Cross-section	Constant cross-section	Continuous fibers usually with polyester and vinylester resins
Hand lay-up	Slow	High	High	Small to large	Simple to complex	Prepreg and fabric with epoxy resin
Wet lay-up	Slow	Medium	Medium to high	Medium to large	Simple to complex	Fabric mat with polyester and Epoxy Resins
Spray-up	Medium to fast	Low	Low	Small to medium	Simple to complex	Short fiber with catalyzed resin
RIM	Medium	Low to medium	Medium	Small to medium	Simple to complex	Preform and fabric with vinylester and epoxy
SRIM	Fast	Low	Medium	Small to medium	Simple to complex	Fabric or preform with polyisocyanurate resin
Compression molding	Fast	Medium	Medium	Small to medium	Simple to complex	Molded compound (e. g. SMC, BMC)
Stamping	Fast	Low	Medium	Medium	Simple to contoured	Fabric impregnated with thermoplastic (tape)
Injection molding	Fast	Low to	Low to medium	Small	Complex	Pallets (short fiber with thermoplastic)
Roll wrapping	Medium to fast	Medium	High	Small to medium	Tubular	Prepregs

There are three types of composite manufacturing processes: open molding, closed molding and cast polymer molding . There are a variety of processing methods within these molding categories, each with its own benefits.

开模成型;闭模成型;浇注聚合物成型

1. Open Molding

Composite materials (resin and fibers) are placed in an open mold, where they cure or harden while exposed to the air. Tooling cost for open molds is often inexpensive , making it possible to use this technique for prototype and short production runs (as shown in Fig. 4－1).

固化;变硬;被暴露
不昂贵
原型件

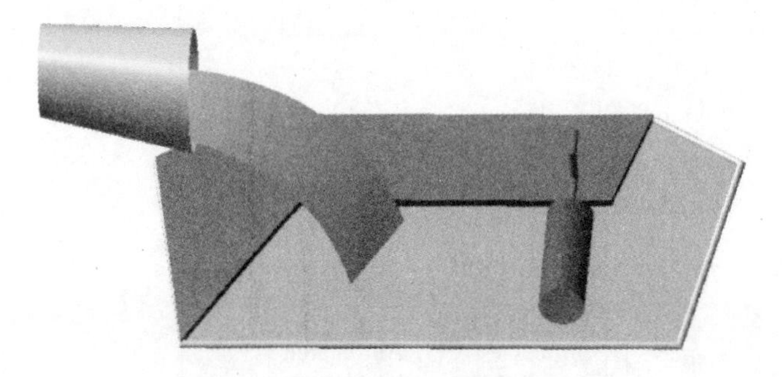

Fig. 4－1 Open molding process

2. Closed Molding

Composite materials are processed and cured inside a vacuum bag or a two-sided mold, closed to the atmosphere . Closed molding may be considered for two cases: first, if a two-sided finish is needed; Second, if high production volumes are required (as shown in Fig. 4－2).

真空袋
模具;大气
情况
高产量

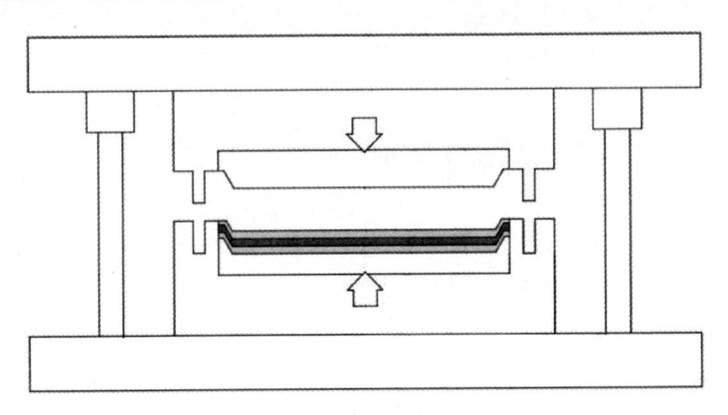

Fig. 4－2 Closed molding process

3. Cast Polymer Molding

A mixture of resin and fillers are poured into a mold (typically without reinforcements) and left to cure or harden. These molding methods sometimes use open molding and sometimes use closed molding (as shown in Fig. 4－3).

填料;倾倒
增强体
方法

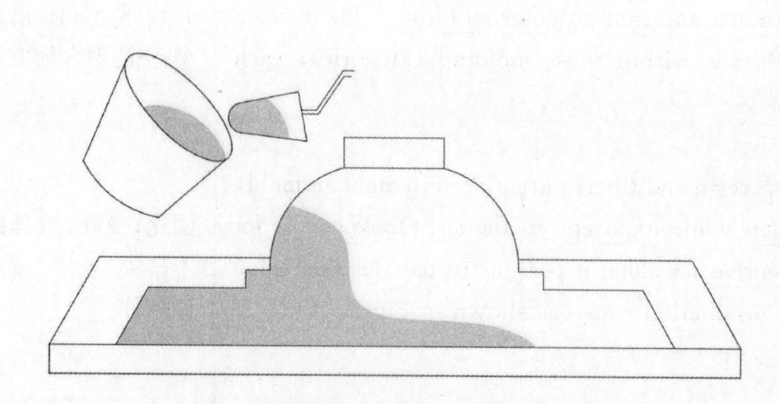

Fig. 4 - 3 Cast polymer molding process

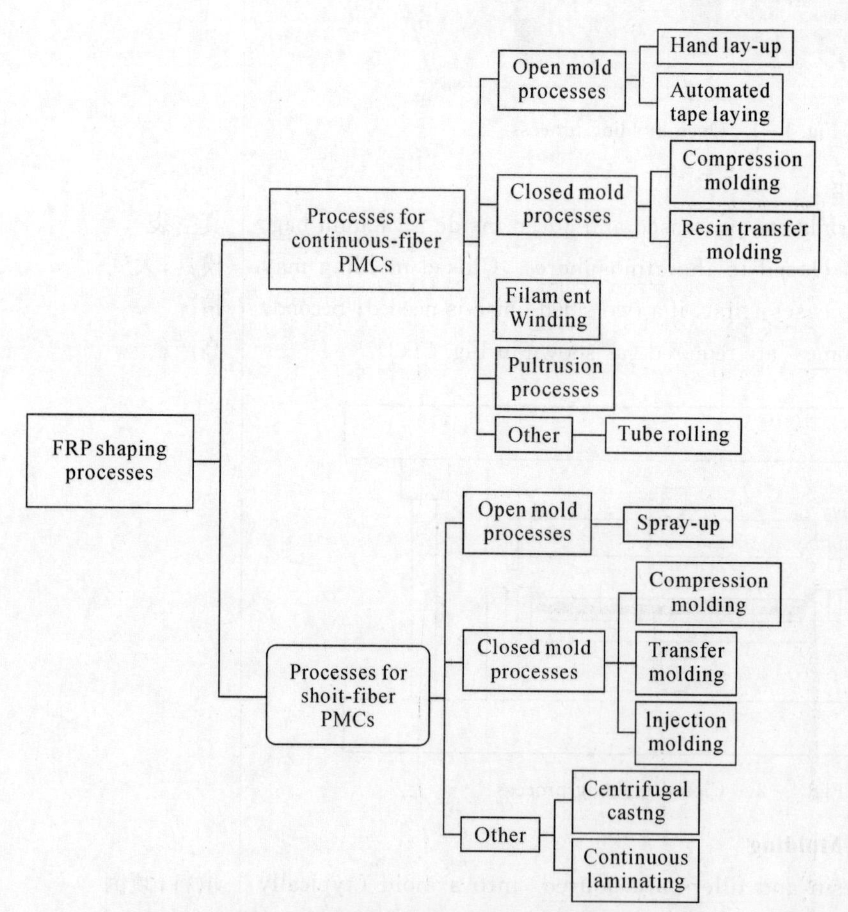

Fig. 4 - 4 The classification of composite molding processes

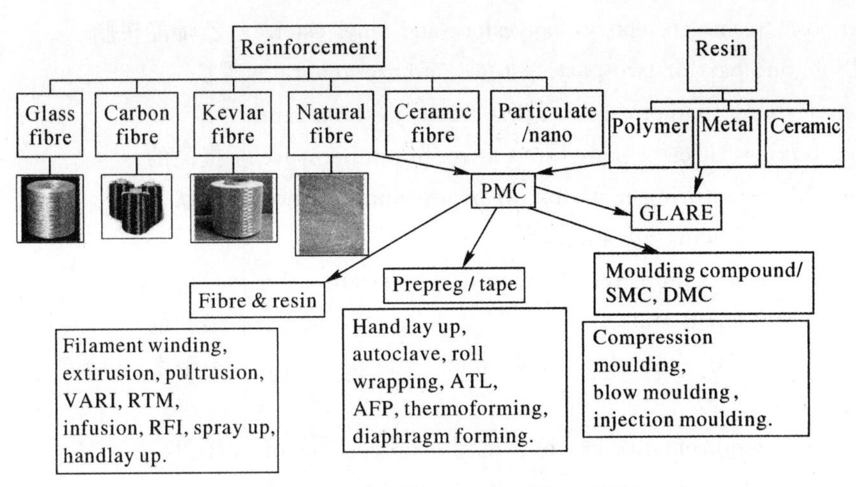

Fig. 4 - 5 Summary of composite manufacturing processes

Unit 1 Thermoset and Thermoplastic Composites Processing

Many thermoplastic products use short discontinuous fibers as a reinforcement. Most commonly fiber is glass, but carbon fiber too. This increases the mechanical properties and is technically considered a fiber reinforced composites, however, the strength is not nearly as comparable to continuous fiber reinforced composites.

不连续的

可比较的

In general, FRP composites refers to the use of reinforcing fibers with a length of 6. 35mm or greater. Recently, thermoplastic resins have been used with continuous fiber creating structural composite products. There are a few distinct advantages and disadvantages of thermoplastic composites and thermoset composites.

明显的

1. Advantages of Thermoset Composites Processing

Thermosets, in basic terms, are materials that undergo a chemical reaction (cure) and transform from a liquid to a solid. In its uncured form, the material has very small, unlinked molecules (known as monomers). The addition of a second material (catalyst) and/or heat or some other activating influence will initiate the chemical reaction. During this reaction the molecules cross-link and form significantly longer molecular chains, causing the material to solidify. This change is permanent and irreversible. Subsequently, exposure to high heat will cause the material to degrade, not melt.

经历
改变;未固化的
单体
催化剂
活性剂、激发

永久的;不可逆的
降解

This is because these materials typically degrade at a temperature below where it would be able to melt.

The common thermoset resins are epoxy, polyester, and vinyl ester. These materials could be one-part or two-part systems and are generally in the liquid state at room temperature. These resin systems are then cured at elevated temperatures or sometimes at room temperature to get the final shape. Manufacturing methods for processing thermoset composites provide the following advantages.

vinyl ester	乙烯酯树脂
elevated	提高的
shape	形状

①Processing of thermoset composites is much easier because the initial resin system is in the liquid state.

②Fibers are easy to wet with thermosets, thus voids and porosities are less.

③Heat and pressure requirements are less in the processing of thermoset composites than thermoplastic composites, thus providing energy savings.

④A simple low-cost tooling system can be used to process thermoset composites.

initial	最初的
voids; porosities	空隙率;孔隙率
pressure	压力
energy	能量
process	过程

2. Disadvantages of Thermoset Composites Processing

①Thermoplastic composites require heavy and strong tooling for processing. Moreover, the cost of tooling is very high in thermoplastic composites manufacturing processes. For example, the tooling cost in the injection molding process is typically more than $50,000, whereas a mandrel for the filament winding process costs less than $500.

②Thermoplastic composites are not easy to process and sometimes require sophisticated equipment to apply heat and pressure.

manufacturing	制造
processes	过程
mandrel; filament winding	芯模;纤维缠绕
sophisticated	复杂的

3. Advantages of Thermoplastic Composites Processing

The initial raw material in thermoplastic composites is in solid state and needs to be melted to obtain the final product. The advantages of processing thermoplastic composites include:

①The process cycle time is usually very short because there is no chemical reaction during processing, and therefore can be used for high-volume production methods. For example, process cycle time for injection molding is less than 1 min and therefore very suitable for automotive-type markets where production rate requirements are usually high.

②Thermoplastic composites can be reshaped and reformed with the application of heat and pressure.

③Thermoplastic composites are easy to recycle.

state; product	状态;产品
cycle time	循环周期
high-volume	高产量
injection	注射
automotive-type	汽车行业
reformed	再成型

4. Disadvantages of Thermoplastic Composites Processing

①Thermoplastic composites require heavy and strong tooling for processing. Moreover, the cost of tooling is very high in thermoplastic composites manufacturing processes. For example, the tooling cost in the injection molding process is typically more than $50,000, whereas a mandrel for the filament winding process costs less than $500. 芯模

② Thermoplastic composites are not easy to process and sometimes require sophisticated equipment to apply heat and pressure.

Unit 2 Manufacturing processing of Composite Materials

Composites manufacturing processes can be broadly subdivided into two main manufacturing categories: manufacturing processes for thermoset composites and manufacturing processes for thermoplastic composites. In terms of commercial applications, thermoset composite parts dominate the composite market. About 75% of all composite products are made from thermoset resins. Thermoset composite processes are much more mature than their thermoplastic counterparts mainly because of the widespread use of thermoset composites as well as its advantages over thermoplastic composite processing techniques. The first use of thermoset composites (glass fiber with unsaturated polyester) occurred in the early 1940s, whereas the use of thermoplastic composites came much later.

细分
商业的
控制
成熟的
不饱和的

The manufacturing processes described in this chapter are discussed under the following headings:

①Basic processing steps.

②Advantages of the process.

③Limitations of the process.

1. Manufacturing Processes for Thermoset Composites

The processing of fiber reinforced laminates can be divided into two main steps: Lay-up and Curing. 铺层;固化

Curing is the drying and hardening (or polymerization) of the resin matrix of a finished composite. This may be done unaided or by applying heat and/or pressure.

硬化
基体;无外援的

Lay-up basically is the process of arranging fiber-reinforced layers (laminae) in a laminate and shaping the laminate to make the part desired (The term "lay-up" is also used to refer to the laminate itself before curing). Unless prepregs are used, lay-up includes the actual creation of laminae by applying resins to fiber reinforcements. Laminate lay-up operations fall into three main groups:

层状板
预浸料

①Molding operations.

②Winding and laying operations.

③Continuous lamination.

Continuous lamination is relatively unimportant compared with quality parameters as not good as other two processes, layers of fabric or mat are passed through a resin dip and brought together between cellophane covering sheets. Laminate thickness and resin content are controlled by squeegee rolls . The lay-up is passed through a heat zone to cure the resin.

层压板
参数;织物
浸润
可剥离纸层
压榨辊

2. Molding Operations

Molding operations are used in making a large number of common composite products. There are two types of processes: open-mold and closed mold.

(1)Open-mold

1)Prepreg lay-up process.

①Introdution.

The hand lay-up process is mainly divided into two major methods: wet lay-up and prepreg lay-up . The wet lay-up process is discussed in previas section. Here, the prepreg lay-up process, which is very common in the aerospace industry, is discussed. It is also called the autoclave processing or vacuum bagging process. Complicated shapes with very high fiber volume fractions can be manufactured using this process. It is an open molding process with low-volume capability . **In this process, prepregs are cut, laid down in the desired fiber orientation on a tool, and then vacuum bagged. After vacuum bagging, the composite with the mold is put inside an oven or autoclave and then heat and pressure are applied for curing and consolidation of the part.**

湿敷;预浸料铺设
热压罐;真空
体积分数
能力
方向
烘箱
共固化

The prepreg lay-up or autoclave process is very labor intensive. Labor costs are 50 - 100 times greater than filament winding, pultrusion, and other high-volume processes; however, for building prototype parts and small quantity runs, the prepreg lay-up process provides advantages over other processes.

原型

②Basic processing steps.

The basic steps in making composite components by prepreg lay-up process are summarized as follows.

a. The prepreg is removed from the refrigerator and is kept at room temperature for thawing .

b. The prepreg is laid on the cutting table and cut to the desired size and orientation.

c. The mold is cleaned and then release agent is applied to the mold surface.

冰箱
解冻
裁剪桌
脱模剂

d. Backing paper from the prepreg is removed and the prepreg is laid on the mold surface in the sequence mentioned in the manufacturing chart.

离型纸
顺序
大纲

e. Entrapped air between prepreg sheets is removed using a squeezing roller after applying each prepreg sheet.

陷入的
挤压辊

f. After applying all the prepreg sheets, vacuum bagging arrangements are made by applying release film, bleeder, barrier film, breather and bagging materials as shown in Fig. 4 - 6.

安排；有孔隔离膜；吸胶层；无孔隔离膜；透气毡

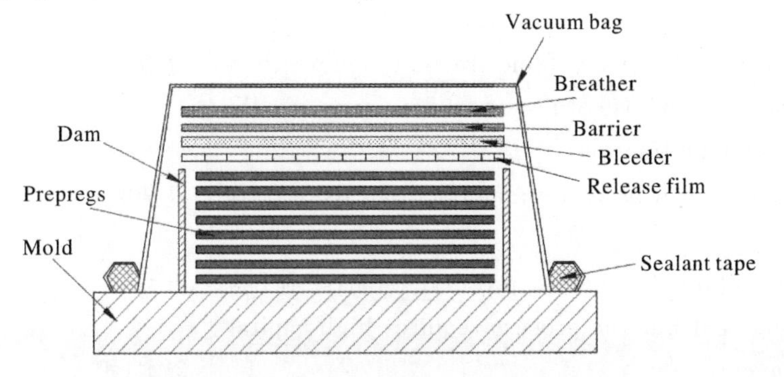

Fig. 4 - 6　Vacuum bagging for prepreg lay-up

g. The entire assembly is then placed into the autoclave using a trolley if the structure is large (as) shown in Fig. 4 - 7.

装配
手推车

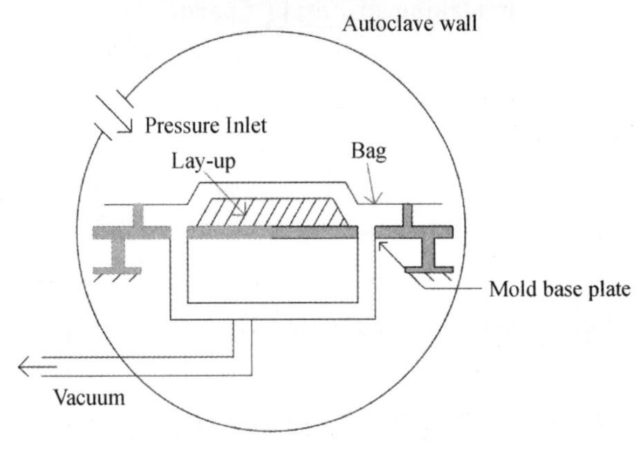

Fig. 4 - 7　Vacuum bagged aerospace part ready to go inside an autoclave.

h. Connections to thermocouples and vacuum hoses are made and the autoclave door is closed. 热电偶;软管

i. The cure cycle data are entered into a computer-controlled machine and followed. 周期

j. After cooling, the vacuum bag is removed and the part is taken out. 冷却

③Advantages of the prepreg lay-up process.

The prepreg lay-up process is very common in the aerospace industry and offers the following advantages:

a. It allows production of high fiber volume fraction (more than 60%) composite parts because of the use of prepregs. Prepregs usually have more than 60% fiber volume fraction. 分数

b. Simple to complex parts can be easily manufactured using this process. 复杂的

c. This process is very suitable for making prototype parts. It has the advantage of low tooling cost but the process requires high capital investment for the autoclave. 原型 资本;投入

d. Very strong and stiff parts can be fabricated using this process. 坚硬的

④Limitations of the prepreg lay-up process.

Although prepreg lay-up is a mature process, it has the following limitations: 成熟的

a. It is very labor intensive and is not suitable for high-volume production applications. 密集的

b. The parts produced by the prepreg lay-up process are expensive. 昂贵的

2) Wet lay-up process.

①Introduction.

In the early days, the wet lay-up process was the dominant fabrication method for the making of composite parts. It is still widely used in the marine industry as well as for making prototype parts. This process is labor intensive and has concerns for styrene emission because of its open mold nature. In this process, liquid resin is applied to the mold and then reinforcement is placed on top. A roller is used to impregnate the fiber with the resin. Another resin and reinforcement layer is applied until a suitable thickness builds up (as) shown in Fig. 4-8. It is a very flexible process that allows the user to optimize the part by placing different types of fabric and mat materials. Because the reinforcement is placed manually, it is also called the hand lay-up process. This process requires little capital investment and expertise and is therefore easy to use. 主要的 海洋的 原型 苯乙烯;释放 浸透 灵活的;使最优化 织物 手动地

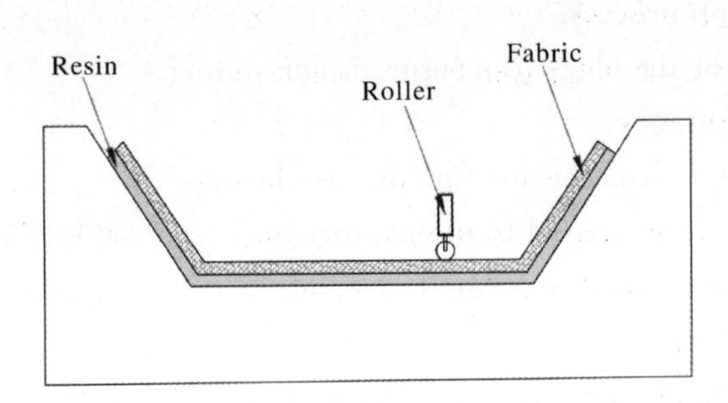

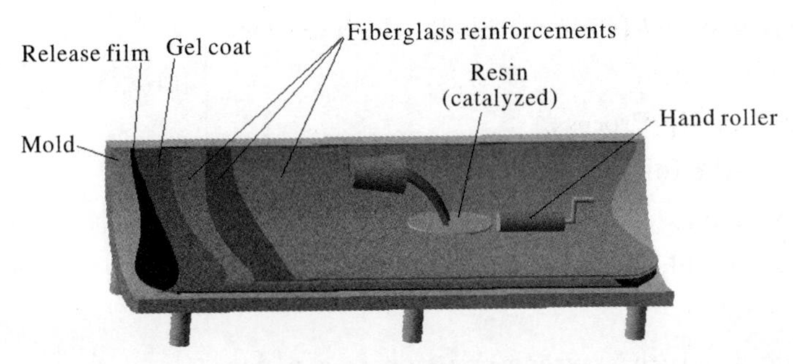

Fig. 4 - 8 Schematic of the wet lay-up process

②Basic processing steps.

The major processing steps in the wet lay-up process include:

a. A release agent is applied to the mold. 脱模剂

b. The gel coat is applied to create a Class A surface finish on the outer surface. The gel coat is hardened before any reinforcing layer is placed. 胶衣

c. The reinforcement layer is placed on the mold surface and then it is impregnated with resin. Sometimes, the wetted fabric is placed directly on the mold surface. 浸透;织物

d. Using a roller, resin is uniformly distributed around the surface. 压滚;均匀地

e. Subsequent reinforcing layers are placed until a suitable thickness is built up. 随后的

f. In the case of sandwich construction, a balsa, foam, or honeycomb core is placed on the laminated skin and then adhesively bonded. Rear-end laminated skin is built similar to how the first laminated skin was built up. 三明治结构;轻木;泡沫蜂窝芯材

g. The partis allowed to cure at room temperature, or at elevated temperature. 升高的

③Advantages of the wet lay-up process.

The wet lay-up process is one of the oldest composite manufacturing techniques with the following advantages:

a. Very low capital investment is required for this process because there is negligible equipment cost as compared to other processes. 可忽略的

b. The process is very simple and versatile. Any type material can be selected with any fiber orientation. 通用的

c. The cost of making a prototype part is low because a simple mold can be used to make the part. In addition, the raw material used for this process is liquid resin, mat and fabric material, which are less expensive than prepreg materials. 标准 毡 预浸料

④Limitations of the Wet Lay-Up Process.

The wet lay-up process has the following limitations:

a. The process is labor intensive.

b. The process is mostly suitable for prototyping as well as for making large structures.

c. Because of its open mold nature, styrene emission is a major concern. 关注点

d. The quality of the part produced is not consistent from part to part.

e. High fiber volume fraction parts cannot be manufactured using this process.

f. The process is not clean.

3) Spray-up process.

①Introduction.

The spray-up process is similar to the wet lay-up process, with the difference being in the method of applying fiber and resin materials onto the mold. The wet lay-up process is labor intensive because reinforcements and resin materials are applied manually. In the spray-up process, a spraygun is used to apply resin and reinforcements with a capacity of 453.6 - 816.5kg material delivered one hour. **In this process a spraygun is used to deposit chopped fiber glass and resin/ catalyst onto the mold. The gun simultaneously chops continuous fiber rovings in a predetermined length (10 - 40mm) and impels it through a resin/ catalyst spray onto the mold.** The spray-up process is much faster than the wet lay-up process and is less expensive choice because it utilizes rovings, which is an inexpensive form of glass fiber (as shown in Fig. 4 - 9).

沉积;催化剂
粗纱
预设定的长度

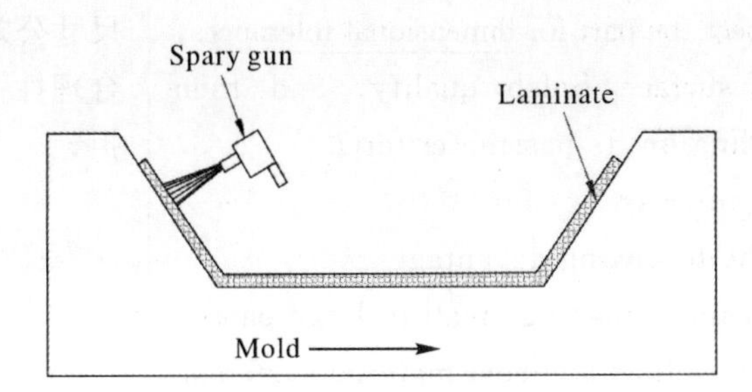

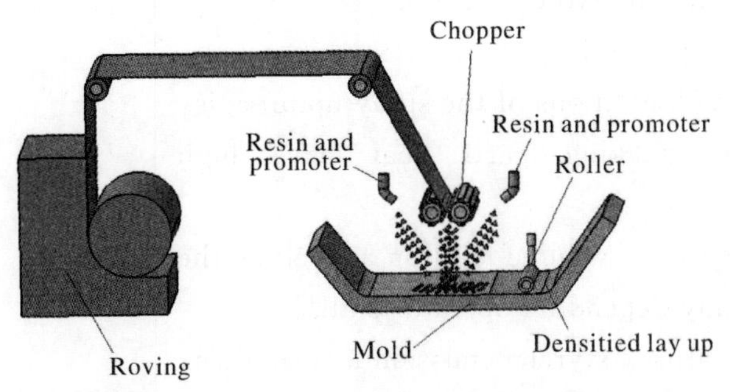

Fig. 4 - 9 Schematic of the spray-up process

②Basic processing steps.

The steps used in the spray-up process are almost the same as for the wet lay-up process, except for the method of creating the laminates. The basic steps are as follows:

a. The mold is waxed and polished for easy demolding.

b. The gel coat is applied to the mold surface and allowed to harden before building any other layer.

c. The barrier coat is applied to avoid fiber print through the gel coat surface.

d. The barrier coat is oven cured. 阻挡层

e. Virgin resin is mixed with fillers such as calcium carbonate or aluminum trihydrate and pumped to a holding tank. 天然的

f. Resin, catalyst, and choppedfibers are sprayed on the mold surface with the help of a hand-held spraygun. The spraygun is moved in a predetermined pattern to create uniform thickness of the laminate (as shown in Fig. 4-9).

g. A roller is used for compaction of sprayed fiber and resin material as well as to create an even and smooth laminate surface. Entrapped air is removed. 压紧

h. Where desirable, wood, foam, or honeycomb cores are embedded into the laminate to create a sandwich structure.

i. The laminate is cured in an oven.

j. The part is demolded and sent forfinishing work.

k. Quality control personnel inspect the part for dimensional tolerances, structural soundness, and good surface finish quality, and then approve or reject the part, depending on its passing criteria. 尺寸公差 合理性 拒绝

③Advantages of the spray-up process.

The spray-up process offers the following advantages:

a. It is a very economical process for making small to large parts.

b. It utilizes low-cost tooling as well as low-cost material systems.

c. It is suitable for small- to medium-volume parts.

④Limitations of the spray-up process.

The following are some of the limitations of the spray-up process:

a. It is not suitable for making parts that have high structural requirements.

b. It is difficult to control the fiber volume fraction as well as the thickness. These parameters highly depend on operator skill.

c. Because of its open mold nature, styrene emission is a concern.

d. The process offers a good surfacefinish on one side and a rough surface finish on the other side. 粗糙的

e. The process is not suitable for parts where dimensional accuracy and process repeatability are prime concerns. The spray-up process does not provide a good surface finish or dimensional control on both or all the sides of the product. 精确度 重复性

4)Pultrusion process.

①Introduction.

The pultrusion process is a low-cost, high-volume manufacturing process in which resin-impregnated fibers are pulled through a die to make the part.

The process is similar to the metal extrusion process, with the difference being that instead of material being pushed through the die in the extrusion process, it is pulled through the die in a pultrusion process. Pultrusion creates parts of constant cross-section and continuous length, as shown in Fig. 4 - 10. 挤压 推出 拉出 等截面 连续长度

②Basic processing steps.

The major steps performed during the pultrusion process are described here. These steps are common in most pultrusion processes:

a. Spools of fiber yarns are kept on creels. 卷轴;纱线;粗纱架

b. Severalfiber yarns from the spool are taken and passed through the resin bath.

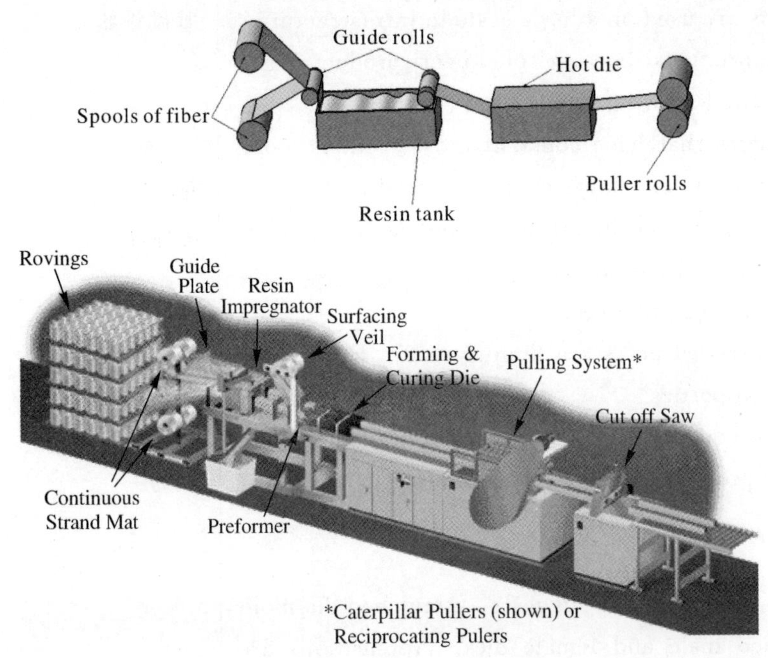

Fig. 4 - 10 Illustration of a pultrusion process

c. Hardener and resin systems are mixed in a container and then poured in the resin bath. 硬化剂;容器

d. The die is heated to a specified temperature for the cure of resin. 模具

e. Resin-impregnatedfibers are then pulled at constant speed from the die, where resin gets compacted and solidified. 不变的 压实的

f. The pultruded part is then cut to the desired length. 拉挤

g. The surface is prepared for painting . Surface preparation is an important element to perform finishing operations because the pultrusion process utilizes internal mold releases . These mold releases are a form of wax that form a film on the outer surface of the part. This film can be removed by solvent wiping, sanding or sandblasting . Solvent wiping is the simplest method of surface preparation. Several solvents (e. g. , toluene, xylene, methylene chloride or acetone) can be used for this purpose. 喷漆 完成 脱模剂 蜡 溶剂;砂纸打磨;喷砂 甲苯二甲苯;二氯甲烷;丙酮

③Advantages of the pultrusion process.

Pultrusion is an automated process with the following advantages: 自动化的

a. It is a continuous process and can be compeletely automated to get the finished part. It is suitable for making high-volume composite parts. Typical production speeds are (0. 61 - 3. 05)m/min.

b. It utilizes low-cost fiber and resin systems and thus provides production of low-cost commercial products. 利用

④Limitations of the pultrusion process.

Pultruded components are used on a large scale in infrastructure, building, and consumer products because of lower product cost. However, pultrusion has the following limitations: 基础设施

a. It is suitable for parts that have constant cross-sections along their length. Tapered and complex shapes cannot be produced. 锥形的

b. Very high-tolerance parts on the inside and outside dimensions cannot be produced using the pultrusion process. 高耐力

c. Thin wall parts cannot be produced.

d. Fiber angles on pultruded parts are limited to 0°. Fabrics are used to get bidirectional properties. 双向的

e. Structures requiring complex loading cannot be produced using this process because the properties are mostly limited to the axial direction. 轴的

(2)Closed-mold

Matched-die molding: As the name suggests, a matched-die mold consists of closely matched male and female dies. Applications are spacecraft part, toys, etc.

Injection molding: The injection process begins with a thermosetting (or sometimes thermoplastic) material outside the mold. The plastic may contain reinforcements or not. It is then forced, under high pressure from a ram or screw, into the cool mold. Applications are auto—parts vanes, engine cowling defrosters and aircraft radomes. 柱塞;螺钉 叶片;防结冰(或霜)装置

1)Resin transfer molding process.

①Introduction.

The resin transfer molding (RTM) process is also known as a liquid transfer molding process. **Although injection molding and compression molding processes have gained popularity as high-volume production methods, their use is mostly limited to nonstructural applications because of the use of molding compounds (short fiber composites). In contrast to these molding processes, the RTM process offers production of cost-effective structural parts in medium-volume quantities using low-cost tooling.** RTM offers the fabrication of near-net-shape complex parts with controlled fiber directions. Continuous fibers are usually used in the RTM process. 普及 非结构的 混合物 近净形

②Basic processing steps.

The steps during the RTM process are summarized below:

a. A thermoset resin and catalyst are placed in tanks A and B of the dispensing equipment. 容器

b. A release agent is applied to the mold for easy removal of the part. Sometimes, a gel coat is applied for good surfacefinish.

c. The preform is placed inside the mold and the mold is clamped. 预成形体

d. The mold is heated to a specified temperature.

e. Mixed resin is injected through inlet ports at selected temperature and pressure. Sometimes, a vacuum is created inside the mold to assist in resin flow as well as to remove air bubbles. 注入;入口 气泡

f. Resin is injected until the mold is completely filled. The vacuum is turned off and the outlet port is closed. The pressure inside the mold is increased to ensure that the remaining porosity is collapsed, (as shown in Fig. 4-11). 出口 孔隙;坍塌

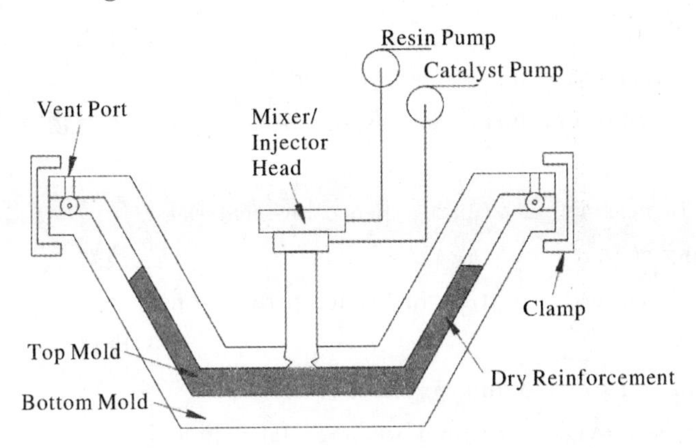

Fig. 4-11 Resin transfer molding (RTM)

g. After curing for a certain time (6 min to 20 min, depending on resin chemistry), the composite part is removed from the mold.

③Advantages of the resin transfer molding process.

Recently, RTM has gained importance in the composites industry because of its potential to make small to large complex structures in a cost-effective manner. RTM provides opportunities to use continuous fibers for the manufacture of structural components in low to medium-volume environments. Some of its major advantages over other composites manufacturing techniques include: 成本有效的 中等产量

a. Initial investment cost is low because of reduced tooling costs and operating expenses as compared to compression molding and injection molding. For this reason, prototypes are easily made for market evaluation. For example, the dish antenna was first made using an RTM process to validate the design features before capital investment was made for compression molding of SMC parts. 最初的 原型 评价;天线 证实

b. Moldings can be manufactured close to dimensional tolerances.

c. RTM processing can make complex parts at intermediate volume 中等的

rates. This feature allows limited production runs in a cost-effective manner. This lends benefits to the automotive market, in which there is a growing need toward lower production volumes per car model and quicker changes to appeal to more niche markets. 吸引;特定客户群

d. RTM provides for the manufacture of parts that have a good surfacefinish on both sides. Sides can have similar or dissimilar surface finishes. 表面粗糙度

e. RTM allows for production of structural parts with selective reinforcement and accurate fiber management. 精确的

f. Higherfiber volume fractions, up to 65%, can be achieved.

g. Inserts can be easily incorporated into moldings and thus allows good joining and assembly features. 连接;装配

h. A wide variety of reinforcement materials can be used.

i. RTM offers low volatile emission during processing because of the closed molding process. 挥发物释放

j. RTM offers production of near-net-shape parts, hence low material wastage and reduced machining cost. 浪费

k. The process can be automated, resulting in higher production rates with less scrap. 废料

④Limitations of the resin transfer molding process.

Although RTM has many advantages compared to other fabrication processes, it also has the following limitations:

a. The manufacture of complex parts requires a good amount of trial-and-error experimentation or flow simulation modeling to make sure that porosity and dry fiber-free parts are manufactured. 反复试验法流量模拟建模;孔隙率

b. Tooling and equipment costs for the RTM process are higher than for hand lay-up and spray-up processes.

c. The tooling design is complex.

2) Vacuum assisted resin transfer molding (VARTM).

The vacuum assisted resin transfer molding (VARTM) process is a closed-mold process that is capable of manufacturing high performance and large-scale fiber reinforced polymer (FRP) parts with low tooling cost. VARTM is an adaptation of the RTM process and is very cost-effective in making large structures such as boat hulls. In this process, tooling costs are cut in half because one-sided tools such as open molds are used to make the part. In this infusion process, fibers are placed in a one-sided mold and a cover, either rigid or flexible, is placed over the top to form a vacuum-tight seal. A vacuum procedure is used to draw the resin into the structure through various types of 辅助的 改进版 船体 灌注 硬的;软的 密封 抽取

ports, (as shown in Fig. 4-12). This process has several advantages compared to the wet lay-up process used in manufacturing boat hulls. Because VARTM is a closed mold process, styrene emissions are close to zero. Moreover, a high fiber volume fraction (70%) is achieved by this process and therefore high structural performance is obtained in the part.

注胶口

苯乙烯释放;零

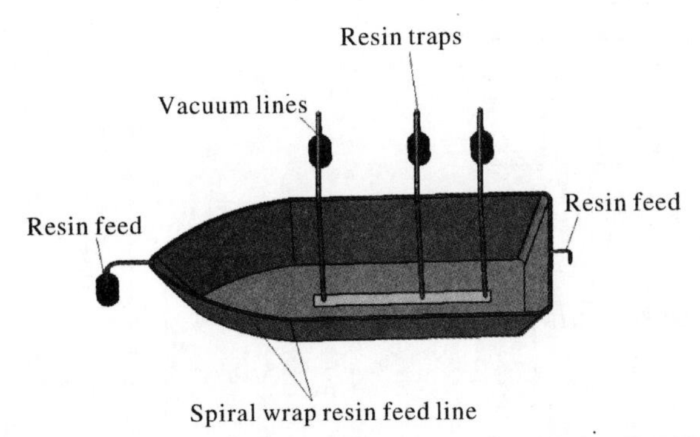

Fig. 4-12 Vacuum assisted resin transfer molding (VARTM)

3)Compression Molding Process.

①Introduction.

Compression molding is very popular in the automotive industry because of its high volume capabilities. This process is used for molding large automotive panels. Sheet molding compounds (SMCs) and bulk molding compounds (BMCs) are the more common raw materials for compression molding. Compression molding is also used for making structural panels using prepregs and core materials, but because of the popularity of compression molding of SMC, the molding of SMC is discussed here.

汽车行业

高产量

面板;片状模塑化合物;块状模塑化合物

预浸料;芯子

②Basic processing steps.

For a clear understanding of the process, the major steps performed during compression molding operations are described below (as shown in Fig. 4-13).

步骤

a. The total volume or weight of the final part is calculated and, based on that, the amount of charge material is determined.

装料

b. The charge material is brought from the storage area and cut to a specific size, for example, a (1×4)in[①]-in. rectangular strip or (6×12) in. rectangular strip. The carrier film from charge material is removed.

储物

矩形条;剥离膜

① 1in=2.54 cm.

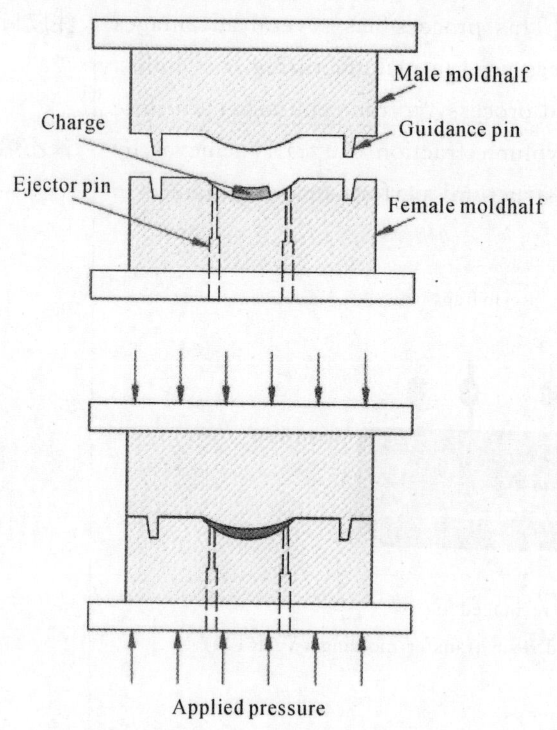

Fig. 4-13　Schematic of the compression molding process.

c. The mold is preheated to about 140℃, or as required by the resin formulation. 配方

d. The charge material is placed on the preheated lower mold half at locations determined by the manufacturing engineer. The charge locations are judiciously selected to get uniform flow across the surface as well as to achieve better mechanical performance. When high surface quality is required, the charge covers a small portion of the mold because flow promotes a good surface finish. When high mechanical performance is desired, the charge covers a larger area to minimize flow induced reinforcement disorientation. 明智的;均匀流动 机械性能 覆盖;比例 提升 最小化;引起 变形

e. The upper half of the mold is closed rapidly at a speed about 40 mm/s. This rapid movement causes the charge to flow rapidly inside the cavity. 上半部分的 腔

f. After about 1 - 4 min of cure cycle, the upper mold half is moved back and thus rele ases the pressure from the mold. 固化周期 释放

g. The part is demolded using ejector pins. 顶杆

③Advantages of the compression molding process.

Compression molding of SMC has become quite a popular process and has established a good name for itself in the automotive industry. Compression molding of SMC provides the following advantages: 建立

a. It offers high-volume production and thus is very suitable for automotive applications. The mold cycle time is only (60 - 240) s.

提供;适合的

b. It offers production of low-cost components at high volume because it utilizes SMC, which is fairly inexpensive.

廉价

c. The process offers high surface quality and good styling possibilities.

样式

d. Multiple parts can be consolidated into one single molded part and thus is very advantageous compared to the metal stamping process.

多部分

金属冲压工艺

e. Today, automotive companies are looking for ways to economically differentiate car and truck models with shorter production runs and more rapid design-to-production schedules. Compressive molding of SMC offers this benefit.

区分;生产线

设计生产时间表

④Limitations of the compression molding process.

Despite the many advantages of compression molding, this process has the following limitations:

a. The initial investment for the process is high because of high equipment and mold costs. However, this initial investment is low compared to sheet metal stamping processes.

片状

b. The process is not suitable for making a small number of parts or for prototyping applications.

c. Compression molding of SMC provides nonstructural parts, but by utilizing ribs and stiffeners, structural parts can be manufactured.

肋骨;加强筋

4)Injection molding.

Injection molding is a manufacturing process widely used for producing items from toys and plastic trinkets to automotive body panels, water bottles, and cell phone cases. A liquid plastic is forced into a mold and cures. It sounds simple, but is a complex process. The liquids used vary from hot glass to a variety of plastics-thermosetting and thermoplastic.

小玩意儿

The first injection molding machine was patented in 1872, and celluloid was used to produce simple everyday items such as hair combs. Just after the Second World War, a much-improved injection molding process "screw injection" was developed and is the most widely used technique today. Its inventor, James Watson Hendry, later developed "blow molding" which is used for example to produce modern plastic bottles.

赛璐珞

螺杆式注塑

吹塑

The plastics used in injection molding are polymers chemicals either thermosetting or thermoplastic. Thermosetting plastics are set by the application of heat or through a catalytic reaction. Once cured, they cannot be remelted and re-used, the curing process is chemical and irreversible. Thermoplastics, however, can be heated, melted and reused. Thermosetting plastics include epoxy, polyester and phenolic resins, while thermoplastics include nylon and polyethylene. **There are almost twenty thousand plastic compounds available for injection molding, which means that there is a perfect solution for almost any molding requirement. Glass is not a polymer, and so it does not fit the accepted definition of thermoplastic though it can be melted and recycled.**

不可逆的

尼龙;聚乙烯

聚合物

3. Winding and Laying Operations

The most important operation in this category is filament winding. Fibers are passed through liquid resin. And then wound onto a mandrel. After lay-up is completed, the composite is cured on the mandrel. The mandrel is then removed by melting, dissolving, breaking-out or some other method.

纤维缠绕

心轴

消溶的

破坏

Filament Winding Process

①Introduction.

Filament winding is a process in which resin-impregnatedfibers are wound over a rotating mandrel at the desired angle. A typical filament winding process (is shown in Fig. 4 - 14 and Fig. 4 - 15), in which a carriage unit moves back and forth and the mandrel rotates at a specified speed. By controlling the motion of the carriage unit and the mandrel, the desire fiber angle is generated. The process is very suitable for making tubular parts. The process can be automated for making high-volume parts in a cost-effective manner. Filament winding is the only manufacturing technique suitable for making certain specialized structures, such as pressure vessels.

旋转的

载运单位;反复地

管状的

容器

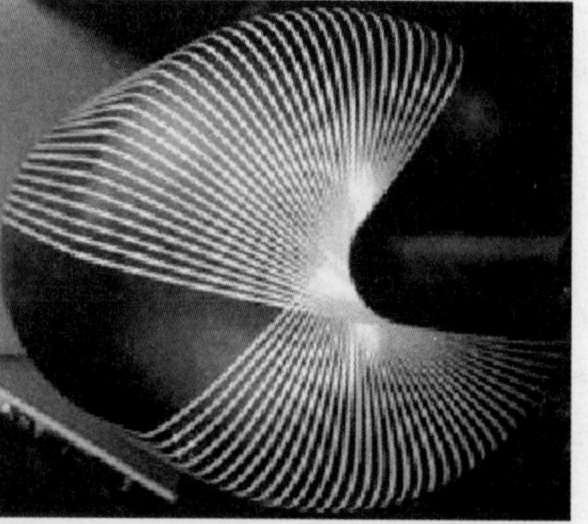

Fig. 4-14 Demonstration of the filament winding operation

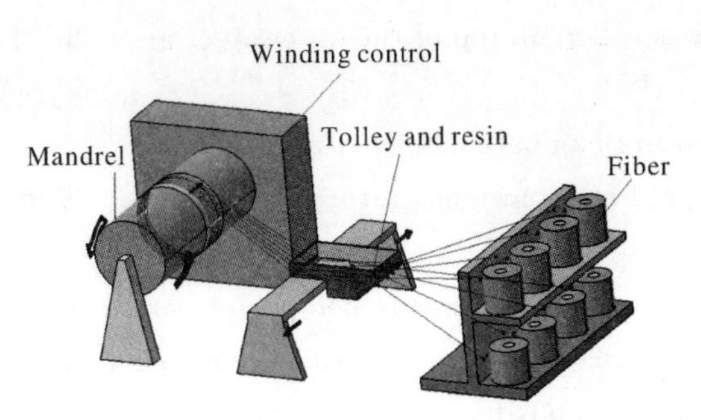

Fig. 4-15 Schematic of the filament winding process

②Basic processing steps.

To understand the entire process more easily, the major steps performed during the filament winding process are described here. These steps are common in all wet filament winding processes:

a. Spools offiber yarns are kept on the creels. 纱线;筒子架

b. Several yarns from spools are taken and passed through guided pins to the payout eye. 放线眼

c. Hardener and resin systems are mixed in a container and then poured into the resin bath. 硬化剂;容器

d. Release agent and gel coat (if applicable) are applied on the mandrel surface and the mandrel is placed between the head and tail stocks of the filament winding machine. 块

e. Resin-impregnatedfibers are pulled from the payout eye and then placed at the starting point on the mandrel surface. Fiber tension is created using a tensioning device. 张力

f. The mandrel and payout eye motions are started. The computer system in the machine creates winding motions to get the desired fiber architecture in the laminate system, as shown in Fig. 4-16. 运动 架构

Fig. 4-16 Demonstration of fiber laydown on a mandrel

g. Fiber bands are laid down on the mandrel surface. The thickness builds up as the winding progresses. 线束

h. To obtain a smooth surfacefinish on the outer surface, a

tefioncoated bleeder or shrink tape is rolled on top of the outer layer after winding is completed. | 聚四氟乙烯吸胶层；（使）缩小

i. The mandrel with the composite laminate is moved to a separate chamber where the composite is cured at room temperature or elevated temperature. | 房间

j. After curing, the mandrel is extracted from the composite part and then reused. For certain applications, the mandrel is not removed and it becomes an integral part of the composite. | 抽取 完整的

③Advantages of the filament winding process.

Filament winding wraps a continuous reinforcement of resin-impregnated fibers onto a mandrel. The combination of mandrel rotation and axial motion of fiber source produce a helical pattern. | 覆盖 螺旋模式

Filament winding has gained significant commercial importance due to its capability in laying down the fibers at a precise angle on the mandrel surface. Filament winding offers the following advantages. | 精确的

a. For certain applications such as pressure vessels and fuel tanks, filament winding is the only method that can be used to make cost-effective and high-performance composite parts. | 燃料槽

b. Filament winding utilizes low-cost raw material systems and low-cost tooling to make cost-effective composite parts. | 利用

c. Filament winding can be automatedfor the production of high-volume composite parts.

④Limitations of the filament winding process.

Filament winding is highly suitable for making simple hollow shapes. However, the process has the following limitations. | 空的

a. **It is limited to producing closed and convex structures. It is not suitable for making open structures such as bathtubs. In some applications, filament winding is used to make open structures such as leaf springs, where the filament wound laminate is cut into two halves and then compression molded.** | 凸面 浴缸 叶片弹簧

b. Not allfiber angles are easily produced during the filament winding process. In general, a geodesic path is preferred for fiber stability. Low fiber angles (0 to 15°) are not easily produced. | 几何的；稳定

c. The maximum fiber volume fraction attainable during this process is only 60%. | 纤维体积分数

d. During the filament winding process, it is difficult to obtain uniform fiber distribution and resin content throughout the thickness of the laminate. | 含量

e. Void content may be high without the use of an autoclave. | 空隙率

In winding thick parts, the process may have to be stopped to allow partial cure of initial layers, as the pressure of additional layers may squeeze out resin.

部分
挤出

Notes

1. Cost concerns have led to several changes in the composites industry. There is a general movement toward the use of less expensive fibers. For example, graphite and aramid fibers have largely supplanted the more costly boron in advanced - fiber composites.

成本问题导致复合材料行业发生了一些变化。产业普遍倾向于使用较便宜的纤维。例如,在先进纤维复合材料中石墨和芳纶纤维很大程度上取代了更昂贵的硼。

2. Lay-up basically is the process of arranging fiber-reinforced layers (laminae) in a laminate and shaping the laminate to make the part desired (The term "lay-up" is also used to refer to the laminate itself before curing.)

铺叠基本上是将纤维增强层(层压板)铺排在层压板中并对层压板进行成形以达到所要求零件的过程(术语"铺叠"也用于指固化前的层压板本身)。

3. Continuous lamination is relatively unimportant compared with quality parameters as not good as other two processes, layers of fabric or mat are passed through a resin dip and brought together between cellophane covering sheets. Laminate thickness and resin content are controlled by squeegee rolls. The lay-up is passed through a heat zone to cure the resin.

由于质量参数没有其他两种工艺好,连续层压工艺相对不重要,在胶膜覆盖层之间织物或毡层通过树脂浸渍并结合在一起。层压板厚度和树脂含量由刮刀辊控制。铺层穿过一个加热区域从而固化树脂。

4. In this process, prepregs are cut, laid down in the desiredfiber orientation on a tool, and then vacuum bagged. After vacuum bagging, the composite with the mold is put inside an oven or autoclave and then heat and pressure are applied for curing and consolidation of the part.

在这个工艺中,预浸料被切割并按所需的纤维方向放置在工作台上,然后抽真空。真空装袋后,将模具和复合材料一起放入烘箱或热压罐中,进行加热和加压从而对零件进行固化。

5. In this process a spraygun is used to deposit chopped fiber glass and resin/catalyst onto the mold. The gun simultaneously chops continuous fiber rovings in a predetermined length (10 - 40 mm) and impels it through a resin/catalyst spray onto the mold.

在这个工艺中,使用喷枪将短切玻璃纤维、树脂和催化剂沉积到模具上。喷枪同时以预定长度(10～40mm)切断连续纤维粗纱线,并推动它通过树脂和催化剂喷射到模具上。

6. The process is similar to the metal extrusion process, with the difference being that instead of material being pushed through the die in the extrusion process, it is pulled through the die in a pultrusion process.

该工艺与金属挤压工艺类似,不同之处在于,在金属挤压工艺中,材料是被推过模具,而拉挤工艺中材料是被拉过模具。

7. Although injection molding and compression molding processes have gained popularity

as high-volume production methods, their use is mostly limited to nonstructural applications because of the use of molding compounds (short fiber composites). In contrast to these molding processes, the RTM process offers production of cost-effective structural parts in medium-volume quantities using low-cost tooling.

虽然注射成型和压缩成型工艺作为大批量生产方法已得到广泛应用,但由于使用了成型化合物(短纤维复合材料),因此它们的使用主要局限于非结构应用。与这些成型工艺相比,RTM 工艺使用低成本工具并以中等批量生产具有成本效益的结构零件。

8. Initial investment cost is low because of reduced tooling costs and operating expenses as compared to compression molding and injection molding. For this reason, prototypes are easily made for market evaluation. For example, the dish antenna was first made using an RTM process to validate the design features before capital investment was made for compression molding of SMC parts.

与压缩成型和注射成型相比,因为降低了模具成本和操作费用,其初始投资成本较低。因此,标准件很容易用于市场评估。例如,在对 SMC 零件的压缩成型进行资本投资之前,首先使用 RTM 工艺制造碟形天线来验证设计特点。

9. The charge material is placed on the preheated lower mold half at locations determined by the manufacturing engineer. The charge locations are judiciously selected to get uniform flow across the surface as well as to achieve better mechanical performance. When high surface quality is required, the charge covers a small portion of the mold because flow promotes a good surface finish. When high mechanical performance is desired, the charge covers a larger area to minimize flow-induced reinforcement disorientation.

在制造工程师确定的位置,将填充材料放置在预热的下半模上。选择合适的填料位置以获得均匀的表面流动,并实现更好的机械性能。当需要高表面质量时,填料只覆盖一小部分,此时流动可以促进良好的表面质量。当需要高机械性能时,填料会覆盖更大的区域,以尽量减少由气流引起的增强体错位。

10. There are almost twenty thousand plastic compounds available for injection molding, which means that there is a perfect solution for almost any molding requirement. Glass is not a polymer, and so it does not fit the accepted definition of thermoplastic-though it can be melted and recycled.

有近 2 万种塑料化合物可用于注塑成型,这意味着对几乎任何成型要求都有一个完美的解决方案。尽管玻璃可以被熔化和回收利用。但它不是聚合物,因为它不符合热塑性塑料的公认定义。

11. It is limited to producing closed and convex structures. It is not suitable for making open structures such as bathtubs. In some applications, filament winding is used to make open structures such as leaf springs, where the filament wound laminate is cut into two halves and then compression molded.

它仅限于生产闭合或凸面结构的产品。它不适合用于生产开放式结构的产品,如浴缸。在某些应用中,缠绕成型用于制造开放式结构产品,如弹簧,在此应用中粗纱线缠绕得到的层压板被切成两半并进行压缩成型。

Exercises

1. Translate the following sentences into Chinese.

(1)As important as savings on materials may be, the real key to cutting composite costs lies in the area of processing.

(2)In general, FRP composites refers to the use of reinforcing fibers with a length of 6.35 mm or greater.

(3)These resin systems are then cured at elevated temperatures or sometimes at room temperature to get the final shape.

(4) Many thermoplastic polymers are addition-type, capable of yielding very long molecular chain lengths (very high molecular weights).

(5) Thermoset composite processes are much more mature than their thermoplastic counterparts mainly because of the widespread use of thermoset composites as well as its advantages over thermoplastic composite processing techniques.

(6)Curing is the drying and hardening (or polymerization) of the resin matrix of a finished composite. This may be done unaided or by applying heat and/or pressure.

(7)Labor costs are (50－100) times greater than filament winding, pultrusion, and other high-volume processes; however, for building prototype parts and small quantity runs, the prepreg lay-up process provides advantages over other processes.

(8)It is a very flexible process that allows the user to optimize the part by placing different types of fabric and mat materials.

(9)The pultrusion process is a low-cost, high-volume manufacturing process in which resin-impregnated fibers are pulled through a die to make the part.

(10) RTM offers the fabrication of near-net-shape complex parts with controlled fiber directions.

(11)In this infusion process, fibers are placed in a one-sided mold and a cover, either rigid or flexible, is placed over the top to form a vacuum-tight seal.

(12)Injection molding is a manufacturing process widely used for producing items from toys and plastic trinkets to automotive body panels, water bottles and cell phone cases.

(13)In winding thick parts, the process may have to be stopped to allow partial cure of initial layers, as the pressure of additional layers may squeeze out resin.

2. Translate the following into English.

熔点

高产量

预浸料铺叠

手糊成型

热压罐成型

真空袋成型

喷射成型

拉挤成型

缠绕成型

树脂转移成型

真空辅助树脂转移成型

注射成型

3. Reading comprehension.

(1)Which form of molding has only one model?

A. contact Molding B. compression Molding

C. molding with Vacuum D. resin Injection Molding

(2)What is the main application of stamp forming?

A. thermoplastic material B. thermosetting material

C. foam composite D. filament

(3)According to the text, which statement is NOT TRUE?

A. Filament winding can fabricate tubes of long length.

B. Stamp forming is only applicable to thermoplastic composites.

C. Molding by foam injection allows the processing of pieces of fairly large dimensions made of polyurethane foam reinforced with glass fibers.

D. Thermoplastic resins can be used to make mechanical components with high temperature resistance.

(4)The mold material can be made of ________.

A. metal B. polymer C. wood D. A, B and C

(5)The blades can also be made by ________ except by vacuum forming.

A. molding processes B. filament winding

C. injection molding D. fabrication processes

(6)As important as savings on materials may be, the real key to cutting composites costs lies in the area of ________.

A. repairing B. assembling C. processing D. inspecting

(7)According to the text, which statement is TRUE?

A. Processing of thermoplastic composites is much easier because the initial resin system is in the liquid state.

B. Fibers are easy to wet with thermosets, thus voids and porosities are less.

C. Heat and pressure requirements are less in the processing of thermoplastic composites than thermoset composites, thus providing energy savings.

D. A simple low-cost tooling system can be used to processthermoplatic composites.

(8)________ composites processing requires a lengthy cure time and thus results in ________ production rate than thermoplastics.

A. thermoset higher B. thermoplastic higher

C. thermoplastic lower D. thermoset lower

(9)Which process has high volume production?

A. compression molding B. wet lay-up

C. prepreg lay-up D. none of them

(10) Which process can fabricate parts with high mechanical properties?

A. spray-up B. compression molding

C. RTM D. filament winding

(11) Which one is not molding operation?

A. filament winding B. spray-up

C. compression molding D. RTM

(12) Who is limited to produce closed and convex structures. It is not suitable for making open structures such as bathtubs.

A. spray-up B. compression molding

C. wet lay-up D. filament winding

(13) After applying all the prepreg sheets, vacuum bagging arrangements are made by applying release film, bleeder, barrier film, ________, and bagging materials.

A. sealant tape B. prepregs C. breather D. dam

(14) Molding operations are used in makinga large number of common composite products. There are two types of processes: ________.

A. open-mold B. closed-mold

C. male mold and female mold D. A and B

(15) Which one is closed-mold process?

A. pultrusion B. vacuum-bag molding

C. compression molding D. autoclave molding

4. Questions.

(1) Why is processing of thermoset composites easier than that of thermoplastic composites?

(2) How do you define an ideal manufacturing process and why?

(3) Why is higher processing temperature required in thermoplastic tape winding as compared to the autoclave or hot press technique?

(4) Write down important processing steps in making a bathtub.

(5) What are the limitations of the prepreg lay-up process.

(6) Write down some of the applications of the prepreg lay-up process.

(7) What are the limitations of the wet lay-up process?

(8) Write down some of the applications of the wet lay-up process.

(9) What are the limitations of the wet lay-up process?

(10) Write down some of the applications offilament winding?

(11) What are the major differences in the die for the thermoset pultrusion process and the metal extrusion process?

(12) What are the major processing steps ina RTM process?

(13) Write down the differences between closed molding and open molding processes.

(14) What are the process selection criteria?

Reading Material

Vacuum Infusion Processing

Vacuum infusion processing is a variation of vacuum bagging in which the resin is introduced into the mold after the vacuum has pulled the bag down and compacted the laminate. Vacuum infusion can produce laminates with a uniform degree of consolidation, producing high strength, lightweight structures. This process uses the same low-cost tooling as open molding and requires minimal equipment. Vacuum infusion offers substantial emissions reduction compared to either open molding or wet lay-up vacuum bagging.

The method is defined as having lower than atmospheric pressure in the mold cavity. The reinforcement and core materials are laid-up dry in the mold by hand, providing the opportunity to precisely position the reinforcement. When the resin is pulled into the mold, the laminate is already compacted; therefore, there is no room for excess resin. Vacuum infusion enables very high resin-to-glass ratios and the mechanical properties of the laminate are superior. Vacuum infusion is suitable to mold very large structures and is considered a low-volume molding process (as shown in Fig. 4-17).

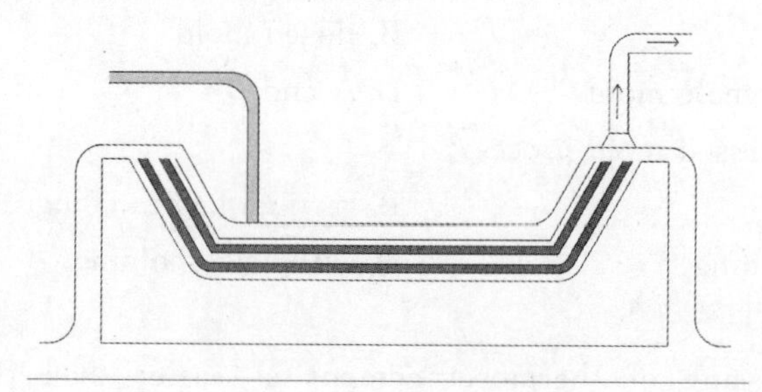

Fig 4-17 Vacuum infusion processing

The mold may be gel coated in the traditional fashion. After the gel coat cures, the dry reinforcement is positioned in the mold. This includes all the plies of the laminate and core material if required. A perforated release film is placed over the dry reinforcement. Next a flow media consisting of a coarse mesh or a "crinkle" ply is positioned, and perforated tubing is positioned as a manifold to distribute resin across the laminate. The vacuum bag is then positioned and sealed at the mold perimeter. A tube is connected between the vacuum bag and the resin container. A vacuum is applied to consolidate the laminate and the resin is pulled into the mold.

New words and Expressions

obstacle	[ˈɔbstəkl]	n. 障碍;障碍物;阻碍
tendency	[ˈtɛndənsi]	n. 倾向,趋势;癖好
aluminum	[əˈlumənəm]	n. 铝
concern	[kənˈsəːn]	v. 关注;关注点;

graphite	[ˈgræfˌaɪt]	n. 石墨;黑铅 v. 用石墨涂
aramid	[ˈærəmɪd]	n. 芳族聚酰胺
supplant	[səˈplænt]	vt. 代替;排挤掉
boron	[ˈborɑn]	n. 硼
monumental	[ˌmɑnjuˈmɛntl]	adj. 不朽的;纪念碑的;非常的
fabricate	[ˈfæbrɪket]	vt. 制造;伪造;装配
criteria	[kraɪˈtɪrɪə]	n. 标准,条件
selection	[sɪˈlɛkʃən]	n. 选择,挑选;选集;精选品
rate	[reit]	n. 比率,率;速度
discontinuous	[dɪskənˈtɪnjuəs]	adj. 不连续的;间断的
comparable	[ˈkɑmpərəbl]	adj. 类似的,可比较的
distinct	[dɪˈstɪŋkt]	adj. 明显的;有区别的
undergo	[ʌndəˈgo]	vt. 经历,经受;忍受
transform	[trænsˈfɔrm]	vt. 改变,使……变形;转换
uncured	[ʌnˈkjʊrd]	adj. 未固化的
catalyst	[ˈkætəlɪst]	n. 催化剂;刺激因素
activate	[ˈæktəˌveit]	v. 使活动;激活;活性化
initiate	[ɪˈnɪʃɪet]	vt. 开始,创始;发起
permanent	[ˈpɝmənənt]	adj. 永久的,永恒的;不变的
irreversible	[ˌɪrɪˈvɝsəbl]	adj. 不可逆的;不能取消的
degrade	[dɪˈgred]	vt. 使……降解 vi. 降级;退化
elevated	[ˈɛlɪvetɪd]	adj. 提高的;高尚的;
shape	[ʃeip]	n. 形状;模型
initial	[ɪˈnɪʃəl]	adj. 最初的;字首的
voids	[vɔɪdz]	n. 空洞,孔洞;空隙率
pressure	[ˈprɛʃə]	n. 压力;压迫
energy	[ˈɛnədʒi]	n. 能量;精力;活力;精神
process	[ˈprəuses]	n. 过程;过程;工艺流程
manufacturing	[ˌmænjuˈfæktʃərɪŋ]	v. 生产,制造;编造
mandrel	[ˈmændrəl]	n. (圆形)心轴
sophisticated	[səˈfɪstɪketɪd]	adj. 复杂的;精致的
commercial	[kəˈmɝʃəl]	adj. 商业的;营利的
dominate	[ˈdɔmineit]	vt. 控制;占优势;在……中占主要地位
mature	[məˈtʃʊr]	adj. 成熟的;充分考虑的
unsaturated	[ʌnˈsætʃəˌreɪtɪd]	adj. 不饱和的
lay-up	[ˈleiʌp]	n. 铺叠
curing	[ˈkjʊərɪŋ]	n. [化学]固化

matrix	[ˈmeɪtrɪks]	n. 基体
laminate	[ˈlæmɪnət]	n. 薄片制品;层压制件
lamination	[læməˈneʃən]	n. 层压;薄板
parameters	[pəˈræmətəz]	n. 参数,参量;界限
fabric	[ˈfæbrɪk]	n. 织物;布
autoclave	[ˈɔtoʊklev]	n. 热压罐
vacuum	[ˈvækjʊəm]	n. 真空;空间
volume	[ˈvɑljum]	n. 量;体积;卷;音量;大量;册
fractions	[ˈfrækʃən]	n. 分数;小部分,片段
capability	[ˌkepəˈbɪlətɪ]	n. 才能,能力;性能,容量
orientation	[orɪɛnˈteiʃən]	n. 方向;定向;适应
oven	[ˈʌvən]	n. 炉,灶;烤炉,烤箱
consolidation	[kən,sɑləˈdeʃən]	n. 巩固;合并;团结
prototype	[ˈprotəˈtaɪp]	n. 原型;标准,模范
refrigerator	[rɪˈfrɪdʒəretə]	n. 冰箱,冷藏库
thawing	[θɔːɪŋ]	n. 融化;熔化
sequence	[ˈsikwəns]	n. 序列;顺序;续发事件
arrangements	[əˈrendʒmənts]	n. 安排;准备;整理
assembly	[əˈsɛmbli]	n. 装配;集会,集合
trolley	[ˈtrɔli]	n. 手推车
thermocouples	[ˈθəːməuˌkʌpəl]	n. 热电偶
cycle	[ˈsaɪkl]	n. 循环;周期
fraction	[ˈfrækʃən]	n. 分数;部分;小部分;稍微
complex	[ˈkɑmplɛks;kəmˈplɛks]	adj. 复杂的;合成的
prototype	[ˈprotəˈtaɪp]	n. 原型;标准,模范
capital	[ˈkæpɪtl]	n. 资金;资本
investment	[ɪnˈvɛstmənt]	n. 投资;投入;封锁
stiff	[stɪf]	adj. 呆板的;坚硬的
mature	[məˈtʃʊr]	adj. 成熟的;充分考虑的
intensive	[ɪnˈtɛnsɪv]	adj. 加强的;集中的
expensive	[ɪkˈspɛnsɪv]	adj. 昂贵的;花钱的
dominant	[ˈdɑmɪnənt]	adj. 显性的;占优势的;支配的
marine	[məˈrin]	n. 海运业;舰队
impregnate	[ɪmˈprɛgnet]	vt. 浸透
flexible	[ˈflɛksəbl]	adj. 灵活的;柔韧的;易弯曲的
optimize	[ˈɑptɪmaɪz]	vt. 使最优化,使完善
fabric	[ˈfæbrɪk]	n. 织物;布;组织;构造;建筑
manually	[ˈmænjʊəli]	adv. 手动地;用手
roller	[ˈrolə]	n. [机]滚轴;辊子

uniformly [junə'fɔrmli] adv. 一致地
subsequent ['sʌbsɪkwənt] adj. 随后的
construction [kən'strʌkʃən] n. 建设;建筑物;解释;造句
balsa ['bɔlsə] n. 热带美洲轻木;轻木
foam [foʊm] n. 泡沫
negligible ['nɛglɪdʒəbl] adj. 微不足道的,可以忽略的
versatile ['vɝsətl] adj. 多才多艺的;通用的
mat [mæt] n. 毡
deposit [dɪ'pɑzɪt] vt. 使沉积;存放
catalyst ['kætəlɪst] n. 催化剂;刺激因素
roving ['rovɪŋ] n. 粗纱,粗纺线
predetermined [ˌpridɪ'tɝmɪnd] adj. 预先确定的
compaction [kəm'pækʃən] n. 压紧;密封
embedded [ɪm'bɛdɪd] adj. 嵌入式的
v. 嵌入
soundness ['saʊndnɪs] n. 合理;稳固;完整
reject ['riːdʒɛkt] vt. 拒绝;排斥;抵制;丢弃
rough [rʌf] adj. 粗糙的;粗略的;未经加工
accuracy ['ækjərəsi] n. 精确度,准确性
repeatability [ri'piːtə'biliti] n. 重复性;[计]可重复性;再现性
spool [spʊl] n. 卷轴,线轴;绕线轮,缠线框
yarn [jɑrn] n. 纱线
creel [kril] n. 粗纱架;经轴架
hardener ['hɑrdnə] n. 硬化剂
container [kən'tenə] n. 容器
die [daɪ] n. 冲模,钢模
constant ['kɑnstənt] adj. 不变的;恒定的
compacted ['kɔmpæktid] adj. 压实的;压紧的
v. 压缩
perform [pə'fɔrm] vt. 执行;完成
releases [rɪ'liːsiz] n. 释放
wax [wæks] n. 蜡;蜡状物
solvent ['sɑlvənt] n. 溶剂
sanding ['sændiŋ] n. 砂纸打磨
v. 撒沙;磨光
sandblasting ['sændˌblæst] n. 喷砂;喷砂处理
v. 对……喷沙
toluene ['tɑljʊˌin] n. 甲苯
xylene ['zaɪlin] n. 二甲苯

methylene chloride 二氯甲烷,亚甲基氯

acetone [ˈæsɪton] n. 丙酮

automated [ɔtəˌmetɪd] adj. 自动化的;机械化的
v. 自动化

utilize [ˈjuːtəlaɪz] vt. 利用

high tolerance 高耐力

bidirectional [ˌbaɪdəˈrɛkʃənl] adj. 双向的;双向作用的

axial [ˈæksɪəl] adj. 轴的;轴向的

tanks [tænks] n. 贮水池;大容器

preform [priˈfɔrm] v. 预先形成
n. 粗加工的成品

inlet [ɪnˌlɛt] n. 入口,进口;插入物

bubbles [ˈbʌbəlz] n. 泡沫;气泡

outlet [ˈaʊtlɛt] n. 出口,排放孔

porosity [pɔˈrɑsəti] n. 有孔性,多孔性

collapsed [kəˈlæpzd] adj. 倒塌的;收缩的

initial [ɪˈnɪʃəl] adj. 最初的

evaluation [ɪˌvæljʊˈeʃən] n. 评价;评估;估价

antenna [ænˈtɛnə] n. 天线;触角,触须

validate [ˈvælɪdet] vt. 证实,验证;确认;使生效

joining [dʒɔɪn] n. 连接,接缝;连接物

assembly [əˈsɛmbli] n. 装配;集会,集合

wastage [ˈwestɪdʒ] n. 损耗;消瘦;衰老

carrier [ˈkærɪə] n. 载体;运送者;带菌者;货架

disorientation [dɪsˈɔrɪɛnteʃən] n. 无序的

cavity [ˈkævəti] n. 腔;洞,凹处

celluloid [ˈsɛljulɔɪd] n. 赛璐珞

mandrel [ˈmændrəl] n. (圆形)心轴

rotating [rəʊˈteɪtɪŋ] adj. 旋转的
v. 旋转,转动

tubular [ˈtubjələ] adj. 管状的

tension [ˈtɛnʃən] n. 张力,拉力

shrink [ʃrɪŋk] v. (使)缩小,(使)收缩

convex [ˈkɑnvɛks] n. 凸面,凸出部分

void [vɔɪd] n. 空洞,孔洞;空隙率

squeeze [skwiz] v. 挤;压榨;使挤进;向……施加压力

hollow [ˈhɔləʊ] adj. 空的;中空的

lie in 在于……;睡懒觉;待产

production rate 生产率

filament winding	纤维缠绕;灯丝绕阻;[电子]灯丝电源绕组
cellophane covering	玻璃纸覆盖
squeeze roll	压榨辊;挤水辊
wet lay-up	湿敷
prepreg lay-up	预浸料铺设
cutting table	裁剪桌;收割台
release agent	脱模剂;隔离剂
backing paper	底子纸;背纸
barrier film	阻挡膜,阻挡层
stray emission	杂散发射
gel coat	胶衣,凝胶涂层;凝胶漆
honeycomb core	蜂窝芯材
barrier coat	屏蔽性涂层,封闭涂层
dimensional tolerance	尺寸公差
near net shape	准精化
volatile mission	挥发物释放
trial-and-error	反复试验法,试错法
ejector pin	出坯杆,起模杆;推顶杆,推钉
screw injection	螺杆式注塑
blow molding	吹塑,吹塑法
filament winding	纤维缠绕;灯丝绕阻[电子]灯丝电源绕组
breaking-out	卸开;喷火
carriage unit	载运单位
back and forth	反复地,来回地
leaf springs	叶片弹簧
helical pattern	螺旋模式
fuel tanks	燃料槽

Chapter 5 Joining of Composite Materials

Warm - up Discussion:

①Do you know what are the commonly used joining methods in the composites industry?

②Do you know what are the advantages of adhesive bonding over mechanical joints?

③ Are you familiar with how many types of adhesives are commonly available?

连接
胶结
机械连接

In any product, there are generally several parts or components joined together to make the complete assembly. For example, there are several thousands of parts in an automobile, a yacht, or an aircraft. The steering system of an automobile has more than 100 parts. Heloval 43-meter luxury yacht from CMN Shipyards is comprised of about 9,000 metallic parts for hull and superstructure, and over 5,000 different types of parts for outfitting (as) shown in Fig. 5 - 1. These parts are interconnected with each other to make the inal product. The purpose of the joint is to transfer loads from one member to another, or to create relative motion between two members. This chapter discusses joints, which create a permanent lock between two members. These joints are primarily used to transfer a load from one member to another.

产品
装配
汽车;游艇
驾驶系统
奢华
金属的部件;船体;
上部构造;供应
永久的

Fig. 5 - 1 Heloval 43 - meter luxury yacht

Airframe structures consist essentially of an assembly of simple elements connected to form a load transmission path. **The elements, which include skins, Stiffeners, frames and spars, form the major parts such as wings, fuselage and empennage. The connections or joints are potentially the weakest points in the airframe, so can determine its structural efficiency. In general, it is desirable to reduce the number and complexity of joints to minimize weight and cost.** A very important advantage of composites construction is the ability to form unitized components, thus minimizing the number of joints required.

载荷传递路径
加强筋;隔框;翼梁
最薄弱的
连接
最小化

However, the design and manufacture of the remaining joints is still a major challenge to produce safe, cost-effective, and efficient structures (as) shown in Fig. 5-2.

设计;制造
成本效益好的

Fig. 5-2 Structural assembly of aircraft

This chapter is concerned with joints used to connect structural elements made of advanced fiber composite laminated, mainly carbon/epoxy, to other composite parts or to metals. Unite 1 and 2 deal, respectively, with bonded and mechanical joints typical of those used in the manufacture of airframe components. Both design and materials aspects are considered. **The aim of this chapter, when discussing design,**

先进纤维复合材料
层压成型
胶黏的

is to outline simple analytic procedures that provide a physical insight into the behavior of joints involving composites. The materials aspects covered will be those essential to the manufacture of sound joints . 分析的;洞察力 连接

Joint types used in airframe construction can be broadly divided into joints that are mechanically fastened using bolts or rivets, adhesively bonded using a polymeric adhesive , or that feature a combination of discontinuity in the load mechanical fastening and adhesive bonding. 螺栓;铆钉 黏合剂

Joints are usually avoided in a structure as good design policy. In any structure, a joint is the weaker area and most failures emanate from joints. Because of this, joints are eliminated by integrating the structure. Joints have the following disadvantages: 产生 整合

① A joint is a source of stress concentration. It creates discontinuity in the load transfer. 应力集中 不连续;载荷传递

②The creation of a joint is a labor - intensive process; a special procedure is followed to make the joint. 劳动密度

③Joints add manufacturing time and cost to the structure. In an ideal product, there is only one part. Fiber - reinforced composites provide the opportunity to create large, complicated parts in one shot and reduce the number of parts in a structure. 纤维增强复合材料

There are two types of joints used in the fabrication of composite products:

①Adhesive bonding.

②Mechanical joints.

Unit 1 Adhesive Bonding

In adhesive bonding, two substrate materials are joined by some type of adhesive (e. g. epoxy, polyurethane or methyl acrylate). The parts that are joined are called substrates or adherends . The loads are transferred mainly by shear on the surfaces of the elements. 胶结;基层 被黏物 剪切力

Various types of bonded joints are shown in Fig. 5 - 3. The most common type of joint is a single lap joint in where the load is transferred from one substrate to another by shear stresses in the adhesive. The joint strength obtained by double lap joint testing is greater because of the absence of normal stresses. For adhesive selection and characterization purposes, single lap joint tests are conducted because single lap joints are very easy to manufacture. The stepped and scarf joints (as) shown in Fig. 单搭接

5 - 3 provide more strength than single lap joints, but machining of stepped or scarf ends is difIcult.

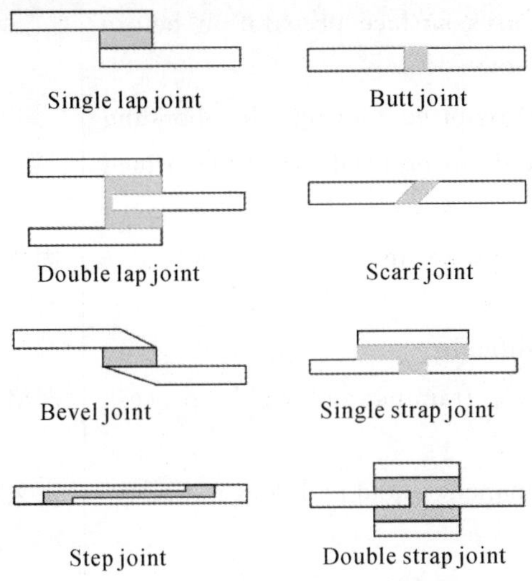

Fig. 5 - 3 Types of adhesively bonded joints

1. Advantages of Adhesive Bonding over Mechanical Joints

Joining of materials using an adhesive offers several beneIts over mechanical joints. In the composites industry, adhesive bonding is much more widely used compared to the metals industry.

黏合剂

①In adhesively bonded joints, the load at the joint interface is distributed over an area rather than concentrated at a point. This results in a more uniform distribution of stresses.

② Adhesively bonded joints are more resistant to flexural, fatigue, and vibrational stresses than mechanical joints because of the uniform stress distribution.

弯曲的

③ The weight penalty is negligible with adhesive bonding compared to mechanical joints.

处罚

④Adhesive not only bonds the two surfaces but also seals the joint. The seal prevents galvanic corrosion between dissimilar adherend materials.

密封;电化学

⑤Adhesive bonding can be more easily adapted to join irregular surfaces than mechanical joints.

不规则的

⑥ Adhesive bonding provides smooth contours and creates virtually no change in part dimensions. This is very important in designing aerodynamic shapes and in creating good part aesthetics.

轮廓

空气动力学的;美学

⑦ Adhesive bonding is often less expensive and faster than mechanical joining.

2. Disadvantages of Adhesive Bonding

Adhesive bonding suffers from the followingdisadvantages:

① Adhesive bonding usually requires surface preparation before bonding.

② Heat and pressure may be required during the bonding operation. This may limit the part size if curing needs to be performed in an oven or autoclave.

③With some adhesives, a long cure time may be needed. 固化周期

④Health and safety could be an issue.

⑤Inspection of a bonded joint is difıcult.

⑥ Adhesive bonding requires more training and rigid process control than mechanical joints. 严格的

⑦Adhesive bonding creates a permanent bond and does not allow repeated assembly and disassembly. 永久的

Unit 2 Mechanical Joints

Mechanical joining is most widely used in joining metal components. Examples of mechanical joints are bolting, riveting, screw, and pin joints. Similar to the mechanical joints of metal components, composite components are also joined using metallic bolts, pins, and screws; except in a few cases where RFI shielding and electrical insulation are required, composite fasteners are used. 螺栓;铆接;螺钉;销

In mechanical joints, loads are transferred between the joint elements by compression on the internal faces of the fastener holes with a smaller component of shear on the outer faces of the elements due to friction. For most mechanical joints, an overlap is required in two mating members and a hole is created at the overlap so that bolts or rivets can be inserted. When screws are used for fastening purposes, mostly metal inserts are used in the composites, the reason being that the threads created in the composites are not strong in shear and therefore metal inserts are used. Fig. 5 - 4 shows examples of bolting and riveting. 摩擦力

In bolted joints, nuts blots, and washers are used to create the joint. In riveting, metal rivets are used. Bolted joints can be a single lap joints, double lap joints, or butt joints, as shown in Fig. 5 - 5. 螺母;螺栓;垫圈 铆钉

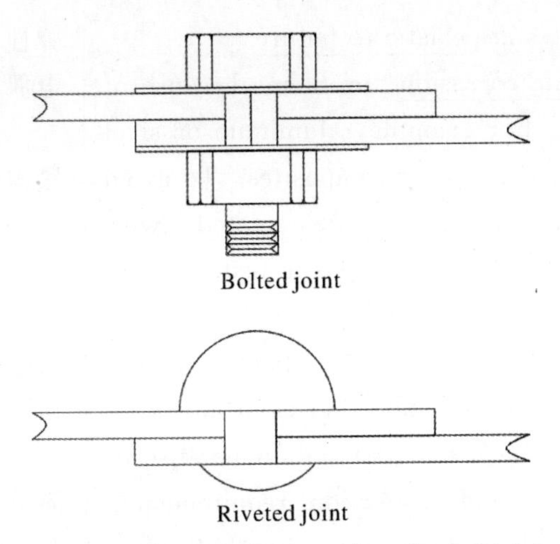

Fig. 5 – 4 Schematic diagram of mechanical joints.

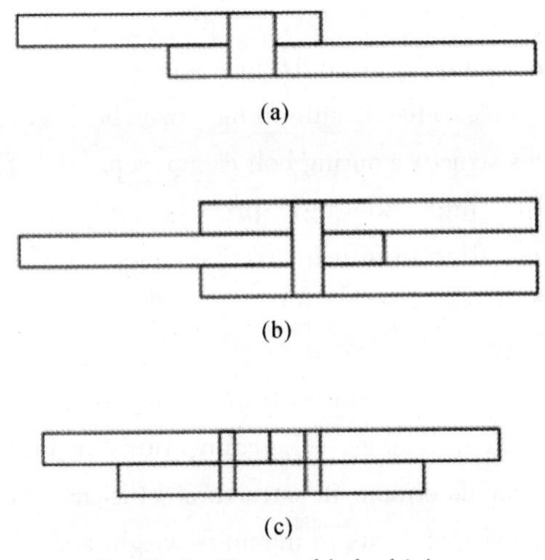

Fig. 5 – 5 Types of bolted joints

(a) single lap joint; (b) double lap joint; (c) butt joint

1. Advantages of Mechanical Joints

①They allow repeated assembly and disassembly for repairs and maintenance without destroying the parent materials.

②They offer easy inspection and quality control.

③They require little or no surface preparation.

2. Disadvantages of Mechanical Joints

①Mechanical joints add weight to the structure and thus minimize the weight – saving potential of composite structures.

②They create stress concentration because of the presence of holes. The composite materials do not have the forgiving characteristics of ductile materials such as aluminum and steel to redistribute local high stresses by yielding. In composites, stress relief

应力集中
宽容的
韧性的
屈服

does not occur because the composites are elastic to failure.

③They create potential galvanic corrosion problems because of the presence of dissimilar materials. For example, aluminum or steel fasteners do not work well with carbon/epoxy composites. To avoid galvanic corrosion, either metal fasteners are coated with nonconductive materials such as a polymer or composite fasteners are used.

④ They createıber discontinuity at the location where a hole is drilled . They also expose ıbers to chemicals and other environments.

Mechanical fastening is usually the lower-cost option because of its simplicity and low-cost tooling and inspection requirements. However, hole-drilling can be highly labor intensive (unless automated) and, if not correctly done, can be highly damaging to the composites.

Joints in aircraft usually require many thousands of expensive fasteners (usually titanium alloy), and extensive shimming may be required to avoid damage to the composites structure during bolt clamp-up. Thus adhesive bonding, despite the high tooling, process and quality control cost , can in many cases offer significant cost savings.

弹性;失效
电偶腐蚀
紧固件
不导电的
不连续
钻孔
简单化;检测
劳动强度
钛合金;填隙
夹紧
质量控制成本

Notes

1. The elements, which include skins, stiffeners, frames and spars, form the major components such as wings, fuselage and empennage. The connections or joints are potentially the weakest points in the airframe so can determine its structural efficiency. In general, it is desirable to reduce the number and complexity of joints to minimize weight and cost.

机翼、机身和尾翼等主要部件由蒙皮、加强筋、框架和翼梁构成。连接或接头可能是机身中最薄弱的部分,由此可以确定其结构效能。一般来说,为了使重量和成本最小化,需要减少接头的数量和复杂性。

2. The aim of this chapter, when discussing design, is to outline simple analytic procedures that provide a physical insight into the behavior of joints involving composites. The materials aspects covered will be those essential to the manufacture of sound joints.

在讨论设计时,本章的目的是概述简单的分析过程,通过此过程可以客观了解复合材料的连接行为。在所涵盖的材料方面,介绍了制造良好接头所必需的材料。

3. Joint types used in airframe construction can be broadly divided into joints that are mechanically fastened using bolts or rivets, adhesively bonded using a polymeric adhesive, or that feature a combination of mechanical fastening and adhesive bonding.

机身结构中使用的接头类型可大致分为使用螺栓或铆钉机械紧固的接头、使用聚合物黏合剂黏合的接头和具有机械紧固或黏合组合的接头。

4. In adhesive bonding, two substrate materials are joined by some type of adhesive

(e.g. epoxy, polyurethane or methyl acrylate). The parts that are joined are called substrates or adherends. The loads are transferred mainly by shear on the surfaces of the elements.

在黏合剂黏合工艺中,两种基材通过某种黏合剂(例如环氧树脂、聚氨酯或丙烯酸甲酯)连接。被连接的部分称为基板或黏合物。荷载主要通过构件表面的剪切力传递。

5. Mechanical joining is most widely used in joining metal components. Examples of mechanical joints are bolting, riveting, screw and pin joints. Similar to the mechanical joints of metal components, composite components are also joined using metallic bolts, pins and screws; except in a few cases where RFI shielding and electrical insulation are required, composite fasteners are used.

机械连接是连接金属部件最广泛使用的方法。使用机械接头的例子有螺栓连接、铆接、螺钉和销接头。与金属部件的机械连接类似,复合部件也使用金属螺栓、销和螺钉连接;少数复合部分由于射频屏蔽和电气绝缘的要求,需要使用复合材料紧固件。

6. In mechanical joints, loads are transferred between the joint elements by compression on the internal faces of the fastener holes with a smaller component of shear on the outer faces of the elements due to friction. For most mechanical joints, an overlap is required in two mating members and a hole is created at the overlap so that bolts or rivets can be inserted.

在机械连接中,载荷在接头元件之间通过紧固件孔内表面上的压缩传递,由于摩擦,紧固件孔外表面上的剪切力分量较小。对于大多数机械连接,两个配合件需要重叠,重叠处会形成一个可以插入螺栓或铆钉的孔。

Exercises

1. Translate the following sentences into Chinese.

(1) However, the design and manufacture of the remaining joints is still a major challenge to produce safe, cost-effective, and efficient structures.

(2) The joint strength obtained by double lap joint testing is greater because of the absence of normal stresses.

(3) Adhesively bonded joints are more resistant to flexural, fatigue, and vibrational stresses than mechanical joints because of the uniform stress distribution.

(4) Mechanical fastening is usually the lower-cost option because of its simplicity and low-cost tooling and inspection requirements.

2. Translate the following into English.

应力集中

机械连接

胶接

单搭接

双搭接

斜接

阶梯接

永久连接

3. Reading comprehension.

(1) ________ is a source of stress concentration.

A. Process B. Fabrication C. Repairing D. Joint

(2) ________ is the type of joints used in the fabrication of composite products.

A. Adhesive bonding B. Mechanical joints

C. A and B D. single lap joint

(3) ________ is very easy to manufacture.

A. Single lap joint B. Butt joint

C. Double lap joint D. Scarf joint

(4) Which is not the advantages of Adhesive bonding over Mechanical joints

A. Uniform distribution of stresses B. Sealing the joint

C. Joining irregular surfaces D. a permanent bond

(5) Which are not a mechanical fastener?

A. Bolts B. Composite fasteners

C. Pins D. Screws

(6) Which is the advantages of Mechanical joints?

A. A permanent joint

B. Requiring surface preparation

C. Being repeated assembly and disassembly

D. Difficult inspection and quality control

4. Questions

(1) What are the commonly used joining methods in the composites industry?

(2) What are the advantages of adhesive bonding over mechanical joints?

(3) How many types of adhesives are commonly available?

(4) How would you select a right adhesive for an application?

(5) Why is there no need of surface preparation in a mechanical joint?

(6) If a joint needs to be designed under peel load, which type of joint will you select and why?

Reading Material

The means of attachment of the stiffeners

The alternative is a stiffened monolithic construction, and here the main issue is the means of attachment of the stiffeners. Some alternatives for attaching stiffeners are shown in Fig. 5 - 6.

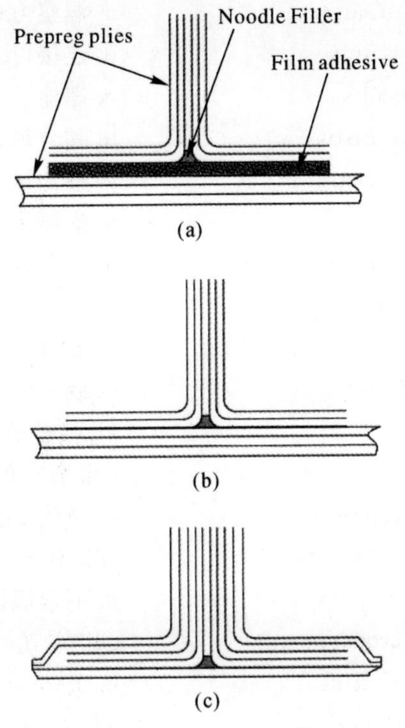

Fig. 5 - 6 Attaching stiffeners

(a) Secondary bonded blade stiffener; (b) Cobonded blade stiffener; (c) integrally cured blade stiffener

From a structural point of view, the integral cocured design is the most effective solution, particularly if the stiffeners must endure buckling of the skins without disboning. however, lay-up costs are higher. To some extent, this cost may be offset by the reduction in parts count. With bonded discrete stiffeners (although cheaper to manufacture), care needs to be taken in matching the stiffness of the panel with the attaching flange, and avoiding excessive through-thickness stresses to avoid the possibility of peel failures. Thorough surface preparation is also essential to ensure a good bond.

New words and Expressions

yacht	[jat]	n. 游艇
		v. 乘游艇
luxury	[ˈlʌkʃərɪ]	n. 奢侈,奢侈品
		adj. 奢侈的
outfitting	[ˈaʊtˌfɪt]	n. 配备;供应(outfit 的 ing 形式)
permanent	[ˈpɝmənənt]	adj. 永久的,不变的
assembly	[əˈsɛmbli]	n. 装配
unitized	[ˈjuːnitaizd]	adj. 联合的;成组的
		v. 统一;单元化
analytic	[ˌænəˈlɪtɪk]	adj. 分析的

insight	[ˈɪnˈsaɪt]	n. 洞察力;洞悉
joints	[dʒɔɪnts]	n. 连接,接头
bolts	[bəʊlts]	n. 螺栓
adhesive	[ədˈhisɪv]	n. 黏合剂;胶带
integrating	[ˈɪntəˌgret]	v. 整合;集成化 n. 集成化;综合化
substrate	[ˈsʌbstret]	n. 基质;基片
scarf	[skarf]	vt. 嵌接
flexura	[ˈfleksjʊrə]	n. 曲折
fatigue	[fəˈtig]	n. 疲劳 vt. 使疲劳 adj. 疲劳的
penalty	[ˈpɛnəlti]	n. 罚款,罚金;处罚
galvanic	[gælˈvænɪk]	adj. 电流的
irregular	[ɪˈrɛgjələ]	n. 不合规格的产品
aerodynamic	[ˌɛrodaɪˈnæmɪk]	adj. 空气动力学的
aesthetics	[ɛsˈθɛtɪks]	n. 美学;美的哲学
rigid	[ˈrɪdʒɪd]	adj. 严格的;坚硬的;精确的
permanent	[ˈpɝmənənt]	adj. 永久的
bolting	[ˈboltɪŋ]	n. 固定(螺栓)的动作
riveting	[ˈrɪvɪtɪŋ]	n. 铆接
screw	[skru]	vt. 旋拧 n. 螺旋;螺丝钉
pin	[pɪn]	vt. 钉住 n. 销
friction	[ˈfrɪkʃən]	n. 摩擦,摩擦力
nuts	[nʌts]	n. 螺母
washers	[ˈwaʃə]	n. 垫圈
forgiving	[fəˈgɪvɪŋ]	adj. 宽大的
ductile	[ˈdʌktaɪl]	adj. 柔软的;易延展的
single lap		单搭接接头
stress concentration		应力集中
adhesive bonding		黏合剂

Chapter 6 Composite Repair

Warm-up Discussion:

①Do you know the major steps to accomplish composite repair tasks?

②Do you know how to describe the honeycomb – structural – composite repair procedure?

Maintenance, repair and operations (MRO) or overhaul involve fixing any sort of mechanical, plumbing or electrical device should become out of order or broken (known as repair, unscheduled, or casualty maintenance. In order to maintain the structural integrity and future operational effectiveness, the repair task will follow the steps in the following flow chart (as shown in Fig. 6 – 1).

大修
水管或电力装置
计划之外的
意外;结构完整性
结构维修手册

Associated with damage inspection, detection and repair following SRM (Structural Repair Manual) instructions, one typical repair procedure of honeycomb core replacement, will be simple but effective described as follow.

Honeycomb Core Replacement Repair Procedure

(1) Inspect the damage.

Thin laminates can be visually inspected and tap tested to map out the damage. **Thicker laminates need more in-depth NDI methods, such as ultrasonic inspection. Check in the vicinity of the damage for entry of water, oil, fuel, dirt, or other foreign matter.** Water can be detected with X-ray, back light, or a moisture detector.

目视;敲击检测
无损检测
附近

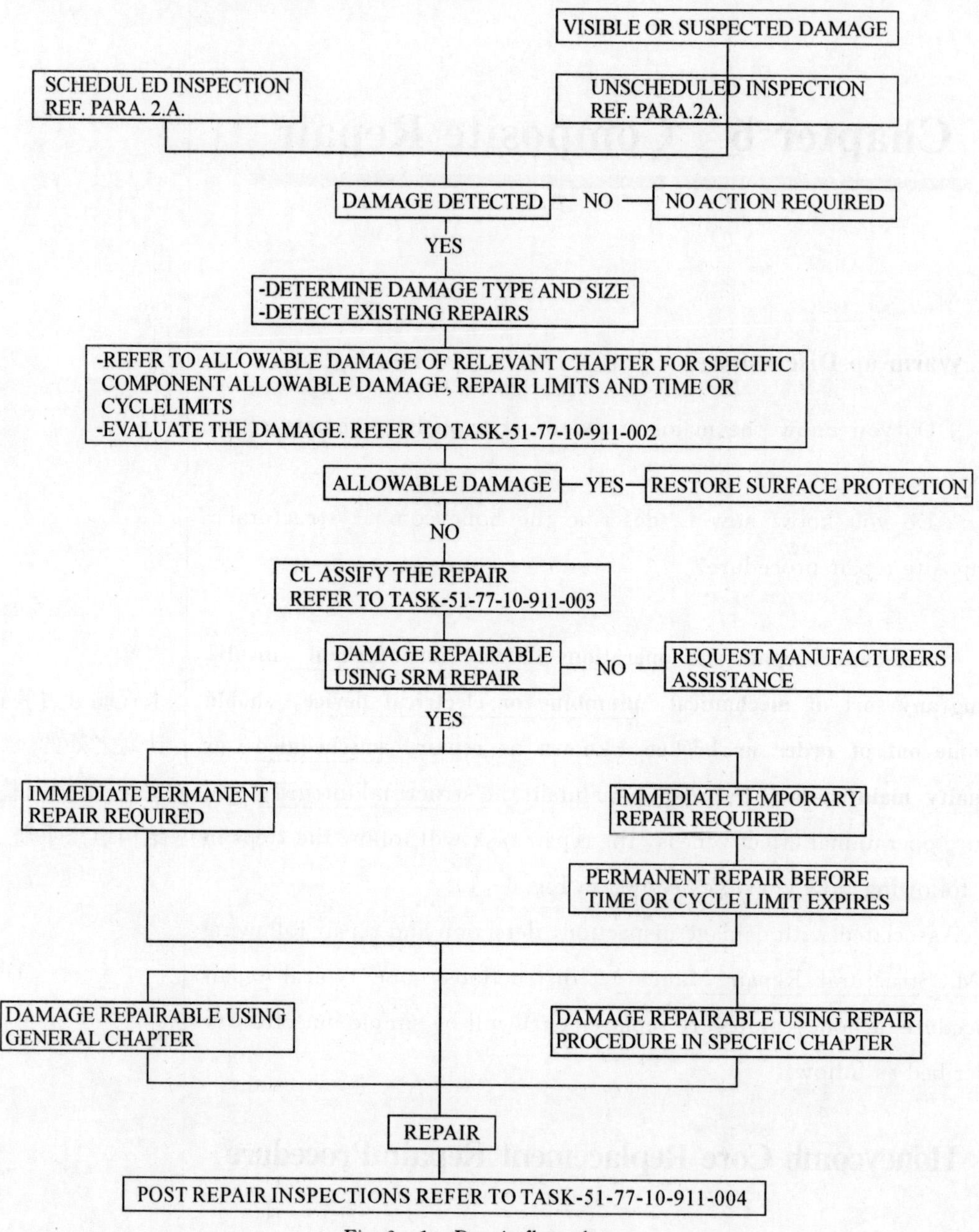

Fig. 6 - 1 Repair flow chart

(2) Remove water from damaged area.

Water needs to be removed from the core before the part is repaired. If the water is not removed, it boils during the elevated temperature cure cycle and the face sheets blow off the core, resulting in more damage. Water in the honeycomb core could also freeze at the low temperatures that exist at high altitudes, which could result in debonding of the face sheets.

沸腾

冻结

脱黏

(3)Remove the damage.

Trim out the damage to the face sheet to a smooth shape with rounded corners, or a circular or oval shape. Do not damage the undamaged plies, core, or surrounding material. If the core is damaged as well, remove the core by trimming to the same outline as the skin.

修整
圆角工具
铺层
轮廓

(4)Prepare the damaged area.

Use a flexible disk sander or a rotating pad sander to taper sand a uniform taper around the cleaned - up damage. Some manufacturers give a taper ratio, such as 30∶1, and others prescribe a taper distance like a 25. 4mm overlap for each existing ply of face sheet. Remove the exterior finish, including conductive coating for an area that is at least 25. 4mm larger than the border of the taper. Remove all sanding dust with dry compressed air and a vacuum cleaner. Use a clean cloth moistened with approved solvent to clean damaged area.

柔性的
斜坡打磨
锥度比;规定
重叠
外层漆层;导电涂层
准许的

(5)Installation of honeycomb core (wet-laying up).

湿铺层

Use a knife to cut the replacement core. The core plug must be of the same type, size, and grade of the original core. **The direction of the core cells should line up with the honeycomb of the surrounding material.** The plug must be trimmed to the right height and be solvent washed with an approved cleaner.

芯格;对齐

(6)Prepare and install the repair plies.

Consult the repair manual for the correct repair material and the number of plies required for the repair. Typically, one more ply than the original number of plies is installed. Cut the plies to the correct size and ply orientation. The repair plies must be installed with the same orientation as that of the original plies being repaired. Impregnate the plies with resin for the wet layup repair, or remove the backing material from the prepreg material. The plies are usually placed using the smallest ply first taper layup sequence.

查阅

(7)Vacuum bagging the repair.

Once the ply materials are in place, vacuum bagging is used to remove air and to pressurize the repair for curing.

加压

(8)Curing the repair.

The repair is cured at required cure cycle. Wet layup repairs can be cured at an elevated temperature up to 65. 5℃ for speeding up the cure. The prepreg repair needs to be cured at an elevated cure cycle. Parts that can be removed from the aircraft could be cured in a hot room, oven, or autoclave. A heating bonder is used for on-aircraft repairs(as shown in Fig. 6 - 2).

固化周期
热补仪

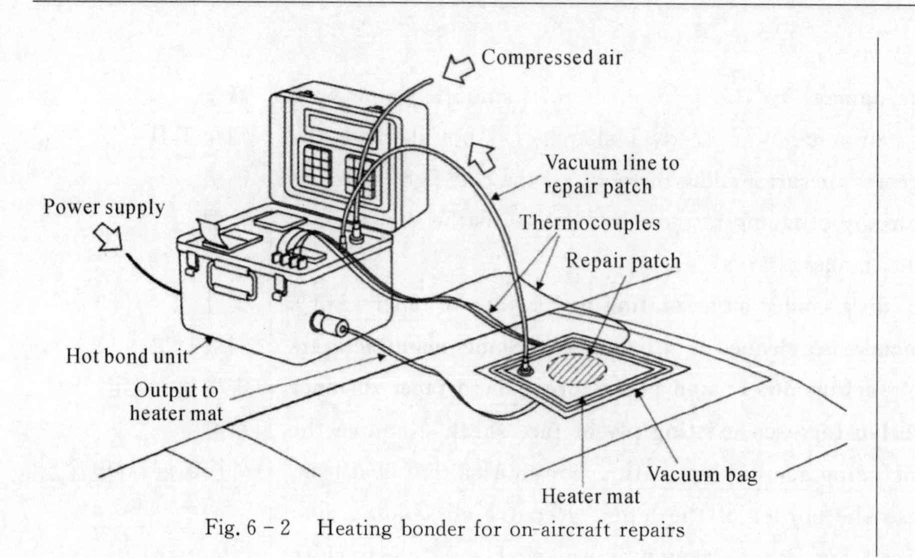

Fig. 6 - 2 Heating bonder for on-aircraft repairs

Notes

1. Maintenance, repair and operations (MRO) or overhaul involve fixing any sort of mechanical, plumbing or electrical device should it become out of order or broken (known as repair, unscheduled or casualty maintenance).

维护、修理、运行(MRO)或大修任何机械、管道或电气设备以防其出现故障或损坏（称为修理、计划外维修、意外事故维修）。

2. Thicker laminates need more in - depth NDI methods, such as ultrasonic inspection. Check in the vicinity of the damage for entry of water, oil, fuel, dirt or other foreign matter.

较厚的层压板需要可探测深度方向的无损检测技术，如超声波检测。需检查损坏处附近是否有水、油、燃料、污垢或其他异物进入。

3. Use a flexible disk sander or a rotating pad sander to taper sand a uniform taper around the cleaned - up damage. Some manufacturers give a taper ratio, such as 30:1, and others prescribe a taper distance like a 25. 4mm overlap for each existing ply of the face sheet.

使用柔性打磨盘或旋转垫片打磨盘对清理后的损坏部位进行均匀的打磨。一些制造商会提供锥度比参数，如 30:1，或规定斜坡宽度，如每一修理层的搭接尺寸为 25. 4mm。

4. The direction of the core cells should line up with the honeycomb of the surrounding material.

芯格的方向应该与周围材料的蜂窝方向一致。

Exercises

1. Fill in the blanks for the following steps.

Although there are many types of composite repair, the most typical repair methods are the wet layup method and the prepreg method, and the general repair procedures are basically similar.

_______ identification.

Indicate the damaged area

Remove damaged materials (such as _______ , honeycomb core, etc.)

______ are made to polish or remove dry repair areas of the surface layer

Cleaning the repair area

Install a new _______ core (if damaged)

_______ the honeycomb core (if damaged)

_______ repairing layers

Check the repair results

_______ the repair area

Repaint the finish

Reading Material

Dark matter composites Ltd course arrangement

Dark Matter Composites Ltd provide the most comprehensive range of composite training courses available. Delegates regularly attend our courses from all industry sectors worldwide which are suitable for individuals and small companies through to tier one suppliers and OEMs. This course is suitable for anyone involved with repairs to composite materials, including repair technicians, supervisors, engineers, designers, researchers and managers as shown in Table 6 - 1.

Table 6 - 1 Course arrangement

	09:00 - 11:00	11:15 - 13:15	13:45 - 15:45	16:00 - 18:00
Monday	Introduction Equipment issue Health & Safety Theory Introduction to composites theory, materials & processes	Theory & Discussion Damage detection Repair of composites overview Theory, Discussion & Demo Dust generation, risk assessment & capture	Theory & Discussion Material removal tools & abrasives Theory, Demo & Practical Non-structural cosmetic & temporary repairs	Theory, Demo & Practical Wet lay-up laminating Matrix preparation, application, curing & exotherm control Application of cosmetic gel coat repairs Application of temporary repairs

(Cont)

	09:00 - 11:00	11:15 - 13:15	13:45 - 15:45	16:00 - 18:00
Tuesday	Demo & Practica Sanding and polishing of cosmetic gel coat repairs Re-filling of cosmetio gel coat repairs Assessment of temporary repairs	Theory, Practical & Discussion Failures modes & predicting failure Impacts & assessment of damage Theory Minor and major structural laminate repairs	Demo & Practical Removal of temporary repair Preparation of tapered/scarf surfaces Preparation of single & double sided repair surfaces	Theory, Demo & Practical Temporary backing structures Process & material matching & preparation
Wednesday	Theory, Demo & Practical Wet lay-up laminating of single & double sided, minor & major structural repairs	Theory & Demg Structural sandwich panel repairs Core preparation & splicing Practical Preparation of type A, B & C structural sandwich panel repair surfaces & materials for infused & pre-preg panels		Theory & Demo Vacuum bagging &materials Caul plates Demo & Practical Wet lay-up & vacuum bagging of type A/B structural repairs
Thursday	Demo & Practical Recognition & orientation of pre-preg materials Laminating of type B/C structural pre-preg sandwich panel repairs	Theory & Practica Curing pre-preg repairs Temperature & pressure profiles Tg points Use of hot bonder for curing pre-preg repairs.	Discussion & Practical Inspection, evaluation and review of repairs completed Destructive testing of repairs	Assessed Practical Assessment of damage Complete a repair strategy Preparation of repairs
Friday	Assessed Practical Completion of repairs to a range of test piece components, including step sanded surface preparation, identification of laminate plies. preparation &application of orientated materials, vacuum bagging, curing, surface finishing, blending and polishing			Reinstate Workshop Written test Equipment return Summary/Feedback

New Word and Expression

honeycomb	[ˈhʌnikoum]	n. 蜂窝
overhaul	[əuvəˈhɔːl]	n. 大修;解体检修
unscheduled	[ʌnˈskedʒuːld]	adj. 事先未安排的
casualty	[ˈkæʒuəlti]	n. 意外事故

vicinity	[vəˈsɪnəti]	n. 邻近,附近
ply	[plaɪ]	n. 板层
outline	[ˈaʊtlaɪn]	n. 轮廓
prescribe	[prɪˈskraɪb]	vi. 规定
overlap	[ˌoʊvərˈlæp]	n. 搭接;重叠的部分
pressurize	[ˈpreʃəraɪz]	vt. 密封;增压
structural integrity		结构完整性
Structural Repair Manual		结构维修手册
visual inspection		目视检查
tap test		敲击测试
face sheet		表皮;面板
trim out		去除;修整
taper sanding		斜坡打磨
taper ratio		锥度比
exterior finish		面漆;外层漆层
conductive coating		导电涂层
compressed air		压缩空气
vacuum cleaner		吸尘器
wet-laying up		湿铺层
core cell		芯格
line up		排列;对齐
ply orientation		铺层角度
heat bonder		热补仪

Chapter 7 Composite Inspection

Warm - up Discussion:

①Do you know the common damage during the aircraft service-life?

②Do you know the definition of non-destructive inspection (NDI) method?

CFRP/GFRP airframe components are nearly immune to chemical deterioration and fatigue while they are primarily damaged by mechanical loads or environmental conditions. With an increased usage of advanced composites in primary and secondary aerospace structural components, it is thus essential to have reliable and repeatable structural inspection and repair procedures to restore damaged composite components. NDI, the non-destructive inspection, will be performed on repaired specimens obtained from aircraft with known service-life history.

免疫
退化
恢复
无损检测
服役周期

1. Damage Identification

Damages in aviation composite materials are mainly classified into two categories: operational damages and manufacturing induced defects. **Common operational damages of composites are because of impact of bird, debris or tool drop during servicing, fatigue stress, atmospheric erosion due to dust or rain and moisture - absorption.** As for manufacturing defects, these types of damages will first occur on a microscopic level , then the microscopic defects will accumulate, resulting in damage on a macroscopic level and repairment needs to be carried out.

诱导的
鸟撞;碎片
侵蚀
微观层面;累积

The following sections will cover different composite damage types and their sources.

(1)De-lamination and de-bond.

This form of composite damage occurs at the interface between the layers in the laminate, along the bond-line between two elements, and

黏结线

between face-sheets and the core of sandwich structures. **De-laminations can form due to stress concentrations at laminate-free edges, matrix cracks, or structural details.** De-laminations may also form from poor processing or from low-energy impact. De-bonds may form similarly. Since de-laminations and de-bonds break the laminate into multiple sub-laminates and reduce the effective stiffness of bonded structural assemblies, they reduce structural stability and strength, posing a safety threat (as shown in Fig. 7-1).

分层;应力集中

脱黏

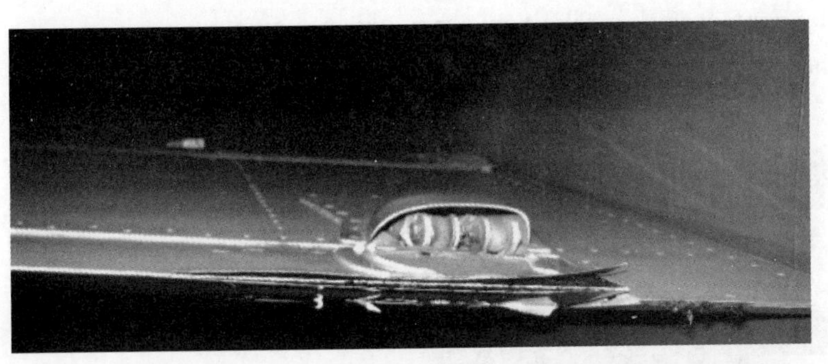

Fig. 7-1 Structural assemblies de-bonding (metal-composite panels)

(2)Cracks.

These types of damage are defined as a fracture of the laminate through the entire thickness (or a portion of the thickness) and involve both fiber breakage and matrix damage. Cracks typically are caused by impact events, but can be the result of excessive local loads (either in the panel or at a fastener hole). In most cases, cracks can be thought of as a combination of fiber breakage and matrix cracking (as shown in Fig. 7-2).

破损

紧固件孔

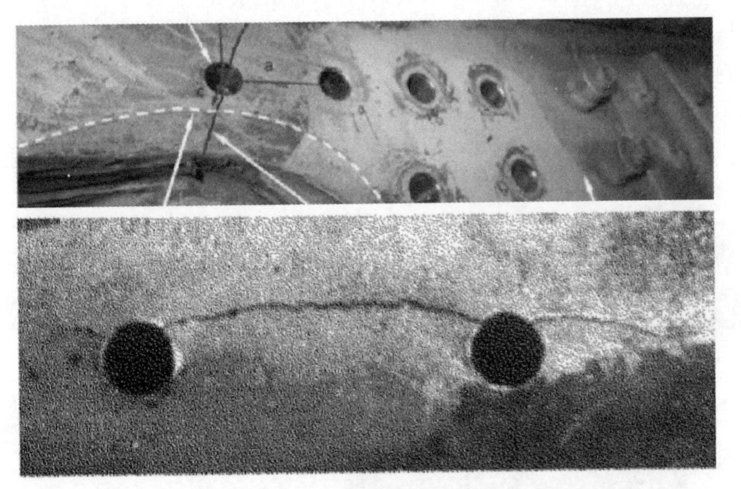

Fig. 7-2 Cracks between fasteners

(3)Scratches, and gouges.

These types of damage are not critical if the damage is limited to the outer layer of resin without any damage to the fibers. If the fibers are damaged, they must be treated as a crack in the affected plies. Unlike metals, composite matrix nicks, scratches and gouges are not likely to grow under repeated loads. A scratch is a line of damage of any depth and length in the material which causes a cross-sectional area change. A sharp object usually causes it. A gouge is a damage area of any size which results in a cross-sectional area change.

缺口;划伤;凿伤

剖面的

It is usually caused by contact with a relatively sharp object which produces a continuous, sharp or smooth groove inside the material (as shown in Fig. 7 - 3).

凹槽

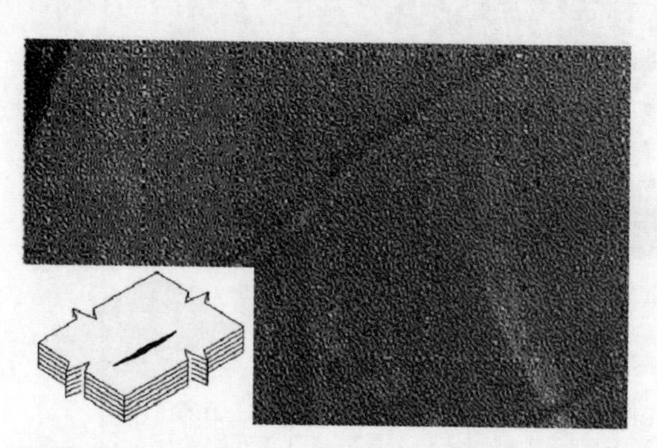

Fig. 7 - 3 Scratch on the composite surface

(4) Dents.

Dents (as shown in Fig. 7 - 4) are typically caused by an impact event. The dent is usually an indication of underlying damage. Damage can consist of one or more of the following: sandwich core damage, face - sheet de - laminations, matrix cracks, fiber breakage, and de - bonds between face - sheets and core. Dents in thin face - sheets of sandwich core often only involve core damage. Dents in solid laminate areas fastened to a substructure (e. g. edge-band areas of sandwich panels) can have associated damage to the substructure.

凹坑

下层的

边缘区域

次级结构;子结构

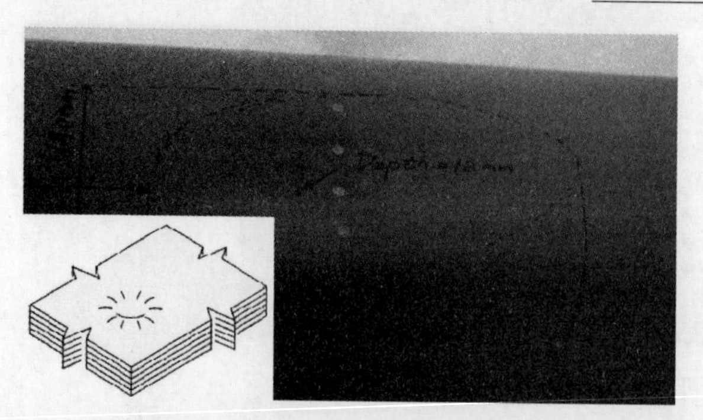

Fig. 7 - 4 Dent on the composite surface

2. Common NDI Test

Per maintenance instructions, damages may be detected using visual inspection or by directed NDI. Visual indications of outside-surface damage should be followed up with a back-side inspection if accessible. **If damage is first detected using visual methods, NDI techniques or even a simple tap hammer, will generally be needed to determine the full extent of the damage and make correct disposition.** It is essential that the proper NDI methods are applied to damage found on composite structure to map the full extent of the damage (as shown in Fig. 7 - 5).

敲击锤
处理方式

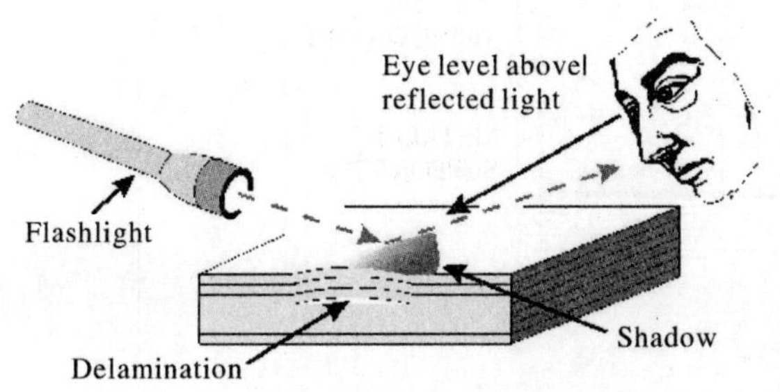

Fig. 7 - 5 Visual inspection method scheme

Many of the damage types described in this section have both visual and hidden damages. Hidden damage in composites usually covers a larger area than visual indications of damage and dominates the residual strength.

隐藏的
主导、决定

(1) Tap testing.

There are many different tap - testing devices ranging from a simple coin tap to woodpecker hammer, where the human ear is used to audibly sense damaged component, automatedly make a recording of changes of sound. Tap testing has been used for damage inspection of composite and metal - bond components. In general, the tap test works well for inspection of damages in thin skin panels, especially useful on sandwich structure with thin face-sheets and honeycomb core. It can work well on solid composite laminate structure if only the near - surface plies are delaminated, but it cannot reliably inspect defects or inside the laminate (as shown in Fig. 7 - 6).

器件
啄木鸟敲击锤
听觉的
可靠的

(2) Ultrasonic inspection.

This type of inspection uses ultrasonic waves transmitted through a part. It basically compares the trace of a standard undamaged laminate of similar thickness with the part being inspected. An inspector using ultrasonic methods must interpret any differences found and, therefore, needs a thorough knowledge of the structure

超声的
轨迹

being inspected. There are generally two types of ultrasonic inspection.

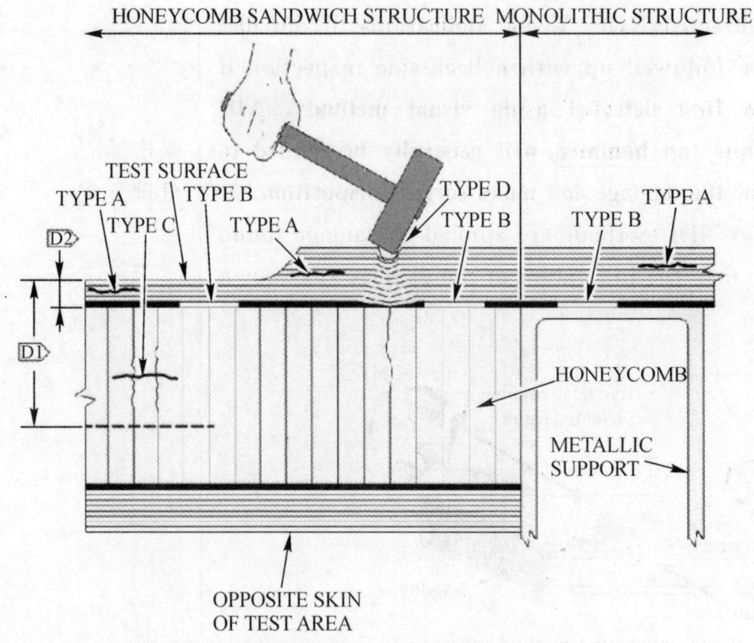

Fig. 7 - 6 Tapping test inspection method scheme

Through-transmission ultrasonics (TTU), which uses two transducers, one to send the ultrasonic wave and one to receive it after traveling through the part. Access to both sides of the part is required. Generally speaking, TTU is applied to repair when the part has been removed from the aircraft.

透射超声
传感器

Pulse-echo (P/E) ultrasonics uses a single transducer and requires access to only one side of the part. It is suitable for different damage detection and post repair inspections (as shown in Fig. 7 - 7).

脉冲反射

Fig. 7 - 7 Ultrasonic inspection equipment (Pulse-echo ultrasonics)

Both TTU and P/E inspection can detect small defects internally bonded elements and plies. Damage through the thickness of a laminate and de-bonds between elements or face-sheets and honeycomb core can be reliably inspected. Special training is usually needed for TTU and P/E inspection since the test standards and a detailed understanding of the part design features, such as damage depth and acoustic impedance parameters, are essential for determining the extent of damage (as shown in Fig. 7 - 8).

声阻抗

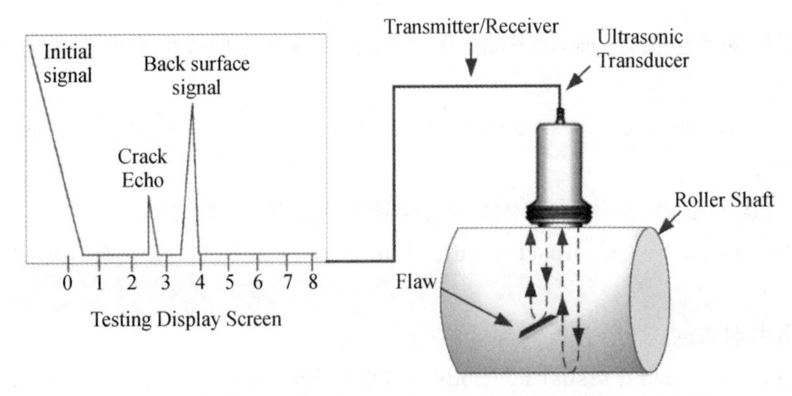

Fig. 7 - 8 Pulse - echo ultrasonics equipment working principle

(3) X-Radiography (X-Ray).

X-ray is often used to detect moisture ingression in honeycomb core of sandwich parts and is sometimes used to detect transverse cracks in laminates. The detailed descriptions of test standards are interpreted in NDT manuals, such as *bond test inspection of metal bonded parts and non - metal laminates*. The use of X-ray on composite parts that are constructed of CFRP is difficult because the absorption characteristics of the fibers and resin are similar and the overall absorption is low. The properties of glass and boron fibers are more suited to the use of X-ray as an inspection method for composites (as shown in Fig. 7 - 9).

进入
横向裂缝
吸波性能

Fig. 7 - 9 X-ray inspection for airplane overhaul check

Notes

1. CFRP/GFRP airframe components are nearly immune to chemical deterioration and fatigue while they are primarily damaged by mechanical loads or environmental conditions.

CFRP/GFRP 机身组件几乎不受化学腐蚀和疲劳效应的影响，载荷或环境老化因素是造成机体损坏的主要因素。

2. Common operational damages of composites are because of impact of bird, debris or tool drop during servicing, fatigue stress, atmospheric erosion due to dust or rain and moisture - absorption.

在役复合材料常见损坏来源包括：鸟撞，碎片、疲劳应力，灰尘或雨水的吸湿老化引起的风蚀效应等。

3. De-laminations can form due to stress concentrations at laminate-free edges, matrix cracks or structural details.

分层损伤，通常是由自由边位置、基体裂纹或微小结构处的应力集中引起的。

4. Dents are typically caused by an impact event. The dent is usually an indication of underlying damage.

凹坑普遍由冲击损伤引起，通常是层合板内部的损伤。

5. If damage is first detected using visual methods, NDI techniques or even a simple tap hammer, will generally be needed to determine the full extent of the damage and make correct disposition.

在目视方法初步检测损伤后，通常还需要运用 NDI 技术甚至是一个简单的敲击锤确定损伤程度并做出正确处理。

6. There are many different tap - testing devices ranging from a simple coin tap to a woodpecker hammer, where the human ear is used to audibly sense damaged component, automatedly make a recording of changes of sound.

敲击测试设备包括简单的敲击币或啄木鸟敲击器，这些设备利用声音判断部件是否损伤，并可自动记录声音的变化。

7. X-ray is often used to detect moisture ingression in honeycomb core of sandwich parts and is sometimes used to detect transverse cracks in laminates.

X 射线法常用于检测夹层结构蜂窝芯的积水损伤，有时也用于探测层合板的横向裂纹。

Exercises

For delaminated areas, Boeing recommends that we use an instrumented Non-Destructive Test (NDT) procedure to find the limits of the delamination. Taking what you have learned in the chapter "Composite Repair" into consideration, discuss the method how to describe the damage areas and design a table as damage recording chart.

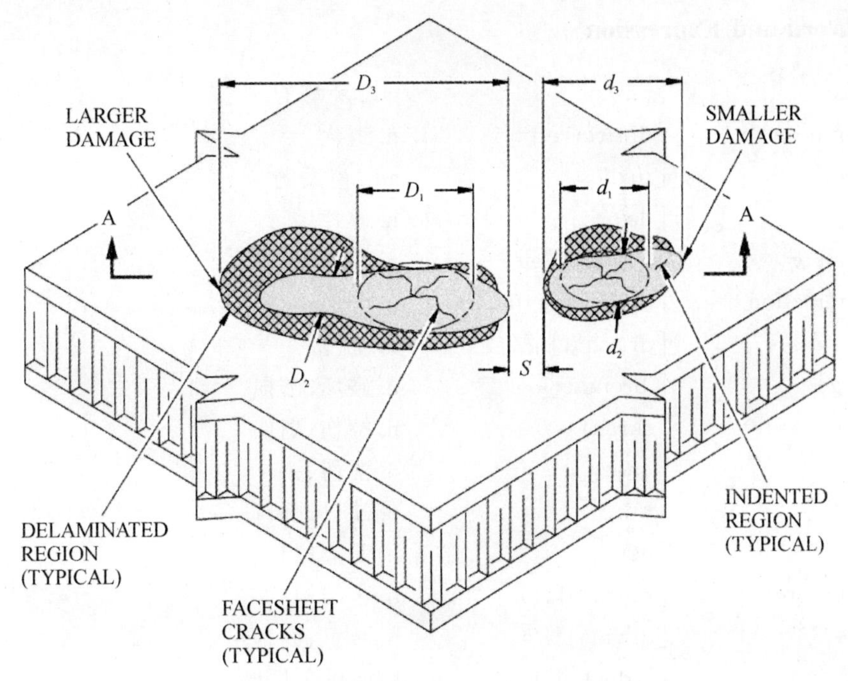

Fig. 7 - 10 Scheme of damaged sandwich panel

Reading Material

As for honeycomb core damage tapping test, the description of possible damage can be arranged into 4 types:

(1) Type A.

De-laminations parallel to the test surface withan minimum extent of $625mm^2$, by a minimum strip width of 20mm or a minimum diameter of 28mm between plies of composite skin with a maximum monolithic laminate thickness of 2mm.

(2) Type B.

Disbond parallel to the test surface with an extent of $625mm^2$ by a minimum strip width of 20mm or a minimum diameter of 28mm between laminated skin and honeycomb core or another structure component with a maximum monolithic skin thickness of 2mm.

(3) Type C.

Honeycomb core damages parallel to the test surface with an extent of $625mm^2$, by a minimum strip width of 20mm, a minimum diameter of 28mm, and in a maximum of (25+5)mm depth depending of skin laminate thickness; skin laminate thickness is 0. 1mm - 0. 8mm +0. 2mm.

(4) Type D.

De-lamination/disbond may occur due to local impact damage (stone impact, dropped tools, lightning strike, localized ground equipment impact, etc.) In these cases, the inspection is only required following visible indications of damage on the surface of the part.

New Word and Expression

immune	[ɪ'mjuːn]	adj. 免疫的
deterioration	[dɪˌtɪriə'reɪʃn]	n. 退化
induced	[ɪn'djʊst]	adj. 诱发的
debris	[də'briː]	n. 碎片
accumulate	[ə'kjuːmjəleɪt]	vi. 累积;积聚
de-lamination	[diˌlæmə'neɪʃən]	n. 分层
de-bond	[diˌbaːnd]	vt. 脱黏
breakage	['breɪkɪdʒ]	n. 破坏;破损
scratch	[skrætʃ]	n. 擦伤;划伤
gouge	[gaʊdʒ]	n. 凿伤
nick	[nɪk]	n. 缺口,裂口
dent	[dent]	n. 凹痕;凹坑
substructure	['sʌbstrʌktʃər]	n. 子结构
devices	[dɪ'vaɪs]	n. 装置;设备
audibly	['ɔːdəbli]	adv. 可听见地
ultrasonic	[ˌʌltrə'saːnɪk]	adj. 超声的
transducers	[træns'duːsər]	n. 传感器
non-destructive		无损的
service-life		在役
damage identification		损伤鉴定
impact of bird		鸟撞
atmospheric erosion		风蚀
microscopic level		微观层面
bond-line		黏结线
stress concentrations		应力集中
fastener hole		紧固件孔
edge-band areas		边缘带区域
tap hammer		敲击锤
woodpecker hammer		啄木鸟敲击锤
through-transmission ultrasonics		透射超声
pulse-echo ultrasonics		脉冲反射波
acoustic impedance		声阻抗
moisture ingression		积水
transverse cracks		横向裂纹

Reference

[1] BAKER A. Composites materials for aircraft structures[M]. Alexander Bell Drive, Reston:American Institute of Aeronautics and Astronautics,2019.

[2] HANCOX N L. Engineering mechanics of composite materials [J]. Mater Design, 1996, 17(2):114。

[3] Hoskin B C, Baker A A. Composite Materials for Aircraft Structure[M]. New York: AIAA Education Series,1986.

[4] BAKER A A. Development and Potential of Advanced Fiber Composites for Aerospace Applications[J]. Materials Forum, 1988,11:27 - 231.

[5] THOSTENSON E T,REN Z, CHOU T W. Advances in the Science and Technology of Carbon Nanotubes and Their Composited: A Review [J]. Composites Science and Technology (UK) 2001,61(13): 1899 - 1912.

[6] CRATCHLEY D,BAKER A A,JACKSON P W. Mechanical Behaviour of a Fibre Reinforced Metal and Its Effect Upon Engineering Applications[J]Metal-Matrix, 1991, 25 (5) :536 - 555.

[7] 郝凌云. 复合材料与工程专业英语[M]. 北京:化学工业出版社,2014.

[8] PEERY D J. Aircraft Structures[M]. NewYork:McGraw - Hill Book Company, 1950.

[9] CONNOLLY J V. Engineering Materials Handbook[J]. The Aeronautical Journal , 1989,93(11):983 - 984.

[10] 刘爱平,林仁伟,陈壁茂. 民用飞机复合材料结构在位修理环境控制方法研究[J]. 航空维修与工程,2021(1):60 - 62.

[11] MOHAMMADI S. A Review on Composite Patch Repairs and the Most Important Parameters Affecting Its Efficiency and Durability[J]. Journal of Reinforced Plastics and Composites,2021, 40(12):3 - 15.

[12] 原志翔,刘礼平. 复合材料损伤检测技术的应用现状[J]. 机械管理开发,2020,35(9):284 - 285,293.

[13] 朱书华,王跃全,童明波. 复合材料层合板阶梯形挖补胶接修理渐进损伤分析[J]. 复合材料学报,2012(6):164 - 169.

[14] 王亮. 浅析复合材料典型修理方法[J]. 科技经济导刊,2019, 691(29):79 - 79.